Innovation Management

Innovation Management

Edited by
Cristian Taylor

Larsen & Keller
www.larsen-keller.com

Innovation Management
Edited by Cristian Taylor
ISBN: 978-1-63549-153-1 (Hardback)

Larsen & Keller
Published by Larsen and Keller Education,
5 Penn Plaza,
19th Floor,
New York, NY 10001, USA

Cataloging-in-Publication Data

Innovation management / edited by Cristian Taylor.
p. cm.
Includes bibliographical references and index.
ISBN 978-1-63549-153-1
1. Technological innovations. 2. Technological innovations--Management.
I. Taylor, Cristian.
T173.8 .I56 2017
600--dc23

Printed and bound in China.

For more information regarding Larsen and Keller Education and its products, please visit the publisher's website www.larsen-keller.com

Table of Contents

Preface

The book aims to shed light on some of the unexplored aspects of innovation management. It will give remarkable insight to the students of this subject. Innovation management is the field which deals with the managing of set goals by using creativity and new ideas, products, etc. The main techniques used under this field are virtual prototyping, brainstorming, project management, portfolio management, product life cycle, etc. This text unfolds the novel aspects of innovation management, which will be crucial for the holistic understanding of the subject matter. The topics covered in this extensive book deal with the core subjects of the field. It also explores all the important aspects of innovation management in the present day scenario. The textbook serves as a valuable source of reference for those interested in innovation management.

Given below is the chapter wise description of the book:

Chapter 1- The management of the innovation processes is known as innovation management. Innovation can depend on existing ideas or completely new ones. The chapter helps in the understanding of the common goal and processes that are involved in the study and in the practice of innovation management.

Chapter 2- New ideas or the development of these ideas into devices and methods is termed as innovation. Some of the topics dealt within this text are open innovation, social innovation, innovation intermediary, pace of innovation and vertical innovation. This section is an overview of the subject matter incorporating all the major aspects of innovation.

Chapter 3- The essential concepts of innovation management include confirmation bias, ambidextrous organization, disruptive innovation, leapfrogging, paradigm shift, eureka effect etc. The ability of an organization to be efficient is known as organizational ambidexterity whereas the tendency of an innovation to disrupt an existing market is known as disruptive innovation. The chapter strategically encompasses and incorporates the key concepts of innovation management, providing a complete understanding.

Chapter 4- The process of understanding the response of the population towards new products is known as the bass diffusion model. Along with the bass model, the theories and models of innovation management explained in this chapter are actor-network theory, diffusion of innovations, induced innovation, social theory, chain-linked model etc.

Chapter 5- This text discusses the methods and techniques of innovation management in a critical manner providing key analysis to the subject matter. Some of the methods explained in section are innovation leadership, creativity techniques, systematic inventive thinking, brainstorming, nominal group technique, social entrepreneurship etc. The chapter serves as a source to understand the major categories relate to innovative management.

Chapter 6- Technological change is the progress made in the production and consumption of technology. Some of the topics elucidated in this text are technological revolution, technological transitions, technological innovation system and technology scouting. This section helps the reader in developing an in-depth understanding of technological changes.

Chapter 7- Research and development is an important constituent of innovation as innovation depends on research and development. The aspects of research and development explained in this text are R&D intensity, knowledge translation, technology brokering and Global R&D management. The aim of this text is to explore the major aspects related to research and development.

Indeed, my job was extremely crucial and challenging as I had to ensure that every chapter is informative and structured in a student-friendly manner. I am thankful for the support provided by my family and colleagues during the completion of this book.

Editor

1

Introduction to Innovation Management

The management of the innovation processes is known as innovation management. Innovation can depend on existing ideas or completely new ones. The chapter helps in the understanding of the common goal and processes that are involved in the study and in the practice of innovation management.

Innovation Management

Innovation management is the management of innovation processes. It refers both to product and organizational innovation.

Innovation management includes a set of tools that allow managers and engineers to cooperate with a common understanding of processes and goals. Innovation management allows the organization to respond to external or internal opportunities, and use its creativity to introduce new ideas, processes or products. It is not relegated to R&D; it involves workers at every level in contributing creatively to a company's product development, manufacturing and marketing.

By utilizing innovation management tools, management can trigger and deploy the creative capabilities of the work force for the continuous development of a company. Common tools include brainstorming, virtual prototyping, product lifecycle management, idea management, TRIZ, Phase–gate model, project management, product line planning and portfolio management. The process can be viewed as an evolutionary integration of organization, technology and market by iterating series of activities: search, select, implement and capture.

Innovation processes can either be *pushed* or *pulled* through development. A *pushed* process is based on existing or newly invented technology, that the organization has access to, and tries to find profitable applications for.

A *pulled* process is based on finding areas where customers needs are not met, and then find solutions to those needs. To succeed with either method, an understanding of both the market and the technical problems are needed. By creating multi-functional development teams, containing both engineers and marketers, both dimensions can be solved.

The product lifecycle of products is getting shorter because of increased competition. This forces companies to reduce the time to market. Innovation managers must therefore decrease development time, without sacrificing quality or meeting the needs of the market.

Innovation Management

Innovation management is based on some of the ideas put forth by the Austrian economist Joseph Schumpeter, working during the 1930s, who identified innovation as a significant factor in economic growth. His book *"Capitalism, Socialism and Democracy"* first fully developed the concept of creative destruction.

Innovation management helps an organization grasp an opportunity and use it to create and introduce new ideas, processes, or products industriously. Creativity is the basis of innovation management; the end goal is a change in services or business process. Innovative ideas are the result of two consecutive steps, *imitation* and *invention*.

By utilizing innovation management tools, management can trigger and deploy the creative capabilities of the work force for the continuous development of a company. Common tools include brainstorming, virtual prototyping, product lifecycle management, idea management, TRIZ, Phase–gate model, project management, product line planning and portfolio management. The process can be viewed as an evolutionary integration of organization, technology, and market, by iterating series of activities: search, select, implement and capture.

Innovation processes can either be *pushed* or *pulled* through development. A *pushed* process is based on existing or newly invented technology that the organization has access to. The goal is to find profitable applications for the already-existing technology. A *pulled* process, by contrast, is based on finding areas where customers' needs are not met and finding solutions to those needs. To succeed with either method, an understanding of both the market and the technical problems are needed. By creating multi-functional development teams, containing both engineers and marketers, both dimensions can be solved.

Innovation, although not sufficient, is a necessary prerequisite for the continued survival and development of enterprises. The most direct way of business innovation is technological innovation and institutional innovation. Management innovation, however, plays a significant role in promoting technological and institutional innovation.

The goal of innovation management within a company is to cultivate a suitable environment to encourage innovation. The suitable environment would help the firms get more cooperation projects, even 'the take-off platform for business ventures'. Senior management's support is crucial to successful innovation; clear direction, endorsement, and support are essential to innovation pursuits.

Managing Complex Innovation

Innovation is a change that outperforms a previous practice. To lead or sustain with innovations, managers need to concentrate heavily on the innovation network, which requires deep understanding of the complexity of innovation. Collaboration is an important source of innovation. Innovations are increasingly brought to the market by networks of firms, selected according to their comparative advantages, and operating in a coordinated manner.

When a technology goes through a major transformation phase and yields a successful innovation, it becomes a great learning experience, not only for the parent industry but other industries as well. Big innovations are generally the outcome of intra- and interdisciplinary networking among technological sectors, along with combination of implicit and explicit knowledge. Networking is required, but network integration is the key to success for complex innovation. Social economic zones, technology corridors, free trade agreements, and technology clusters are some of the ways to encourage organizational networking and cross-functional innovations.

2

Innovation: An Overview

New ideas or the development of these ideas into devices and methods is termed as innovation. Some of the topics dealt within this text are open innovation, social innovation, innovation intermediary, pace of innovation and vertical innovation. This section is an overview of the subject matter incorporating all the major aspects of innovation.

Innovation

Innovation is defined simply as a "new idea, device, or method". However, innovation is often also viewed as the application of better solutions that meet new requirements, unarticulated needs, or existing market needs. This is accomplished through more-effective products, processes, services, technologies, or business models that are readily available to markets, governments and society. The term "innovation" can be defined as something original and more effective and, as a consequence, new, that "breaks into" the market or society. It is related to, but not the same as, invention.

While a novel device is often described as an innovation, in economics, management science, and other fields of practice and analysis, innovation is generally considered to be the result of a process that brings together various novel ideas in a way that they affect society. In industrial economics, innovations are created and found empirically from services to meet the growing consumer demand.

Definition

In a survey of literature on innovation, Edison et al. found over 40 definitions. They also performed an industrial survey to capture how innovation is defined in the software industry. After analysis of the existing definitions whether these definitions comprehensively cover all the dimensions of innovation, they found the following definition to be the most complete:

Innovation is: production or adoption, assimilation, and exploitation of a value-added novelty in economic and social spheres; renewal and enlargement of products, services, and markets; development of new methods of production; and establishment of new management systems. It is both a process and an outcome.

This definition was given by Crossan and Apaydin and it builds on the OECD manual's definition.

Edison et al. also found two interesting dimensions of innovation including: degree of novelty (i.e. whether an innovation is new to the firm, new to the market, new to the industry, and new to the world) and type of innovation (whether it is process or product/ service innovation).

Inter-disciplinary Views

Business and Economics

In business and economics, innovation can be a catalyst to growth. With rapid advancements in transportation and communications over the past few decades, the old world concepts of factor endowments and comparative advantage which focused on an area's unique inputs are outmoded for today's global economy. Economist Joseph Schumpeter, who contributed greatly to the study of innovation economics, argued that industries must incessantly revolutionize the economic structure from within, that is innovate with better or more effective processes and products, as well as market distribution, such as the connection from the craft shop to factory. He famously asserted that "creative destruction is the essential fact about capitalism". In addition, entrepreneurs continuously look for better ways to satisfy their consumer base with improved quality, durability, service, and price which come to fruition in innovation with advanced technologies and organizational strategies.

One prime example was the explosive boom of Silicon Valley startups out of the Stanford Industrial Park. In 1957, dissatisfied employees of Shockley Semiconductor, the company of Nobel laureate and co-inventor of the transistor William Shockley, left to form an independent firm, Fairchild Semiconductor. After several years, Fairchild developed into a formidable presence in the sector. Eventually, these founders left to start their own companies based on their own, unique, latest ideas, and then leading employees started their own firms. Over the next 20 years, this snowball process launched the momentous startup company explosion of information technology firms. Essentially, Silicon Valley began as 65 new enterprises born out of Shockley's eight former employees. Since then, hubs of innovation have sprung up globally with similar metonyms, including Silicon Alley encompassing New York City.

Organizations

In the organizational context, innovation may be linked to positive changes in efficiency, productivity, quality, competitiveness, and market share. However, recent research findings highlight the complementary role of organizational culture in enabling organizations to translate innovative activity into tangible performance improvements. Organizations can also improve profits and performance by providing work groups opportunities and resources to innovate, in addition to employee's core job tasks. Peter Drucker wrote:

Innovation is the specific function of entrepreneurship, whether in an existing business, a public service institution, or a new venture started by a lone individual in the family kitchen. It is the means by which the entrepreneur either creates new wealth-producing resources or endows existing resources with enhanced potential for creating wealth.

–Drucker

According to Clayton Christensen, disruptive innovation is the key to future success in business. The organisation requires a proper structure in order to retain competitive advantage. It is necessary to create and nurture an environment of innovation. Executives and managers need to break away from traditional ways of thinking and use change to their advantage. It is a time of risk but even greater opportunity. The world of work is changing with the increase in the use of technology and both companies and businesses are becoming increasingly competitive. Companies will have to downsize and re-engineer their operations to remain competitive. This will affect employment as businesses will be forced to reduce the number of people employed while accomplishing the same amount of work if not more.

All organizations can innovate, including for example hospitals, universities, and local governments. For instance, former Mayor Martin O'Malley pushed the City of Baltimore to use CitiStat, a performance-measurement data and management system that allows city officials to maintain statistics on crime trends to condition of potholes. This system aids in better evaluation of policies and procedures with accountability and efficiency in terms of time and money. In its first year, CitiStat saved the city $13.2 million. Even mass transit systems have innovated with hybrid bus fleets to real-time tracking at bus stands. In addition, the growing use of mobile data terminals in vehicles, that serve as communication hubs between vehicles and a control center, automatically send data on location, passenger counts, engine performance, mileage and other information. This tool helps to deliver and manage transportation systems.

Still other innovative strategies include hospitals digitizing medical information in electronic medical records. For example, the U.S. Department of Housing and Urban Development's HOPE VI initiatives turned severely distressed public housing in urban areas into revitalized, mixed-income environments; the Harlem Children's Zone used a community-based approach to educate local area children; and the Environmental Protection Agency's brownfield grants facilitates turning over brownfields for environmental protection, green spaces, community and commercial development.

Sources of Innovation

There are several sources of innovation. It can occur as a result of a focus effort by a range of different agents, by chance, or as a result of a major system failure.

According to Peter F. Drucker, the general sources of innovations are different changes in industry structure, in market structure, in local and global demographics, in human perception, mood and meaning, in the amount of already available scientific knowledge, etc.

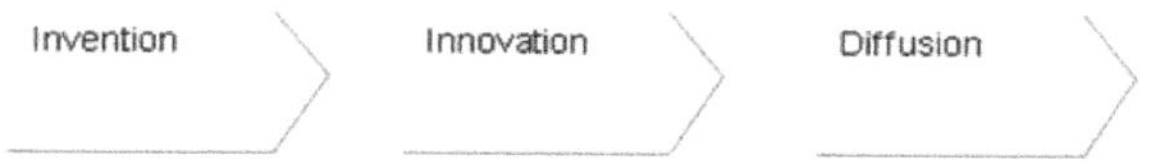

Original model of three phases of the process of Technological Change

In the simplest linear model of innovation the traditionally recognized source is *manufacturer innovation*. This is where an agent (person or business) innovates in order to sell the innovation.

Another source of innovation, only now becoming widely recognized, is *end-user innovation*. This is where an agent (person or company) develops an innovation for their own (personal or in-house) use because existing products do not meet their needs. MIT economist Eric von Hippel has identified end-user innovation as, by far, the most important and critical in his classic book on the subject, *The Sources of Innovation*.

The robotics engineer Joseph F. Engelberger asserts that innovations require only three things:

1. A recognized need,
2. Competent people with relevant technology, and
3. Financial support.

However, innovation processes usually involve: identifying customer needs, macro and meso trends, developing competences, and finding financial support.

The Kline chain-linked model of innovation places emphasis on potential market needs as drivers of the innovation process, and describes the complex and often iterative feedback loops between marketing, design, manufacturing, and R&D.

Innovation by businesses is achieved in many ways, with much attention now given to formal research and development (R&D) for "breakthrough innovations". R&D help spur on patents and other scientific innovations that leads to productive growth in such areas as industry, medicine, engineering, and government. Yet, innovations can be developed by less formal on-the-job modifications of practice, through exchange and combination of professional experience and by many other routes. The more radical and revolutionary innovations tend to emerge from R&D, while more incremental innovations may emerge from practice – but there are many exceptions to each of these trends.

Information technology and changing business processes and management style can produce a work climate favorable to innovation. For example, the software tool company Atlassian conducts quarterly "ShipIt Days" in which employees may work on anything related to the company's products. Google employees work on self directed projects for 20% of their time (known as Innovation Time Off). Both companies cite these bottom-up processes as major sources for new products and features.

An important innovation factor includes customers buying products or using services. As a result, firms may incorporate users in focus groups (user centred approach), work closely with so called lead users (lead user approach) or users might adapt their products themselves. The lead user method focuses on idea generation based on leading users to develop breakthrough innovations. U-STIR, a project to innovate Europe's surface transportation system, employs such workshops. Regarding this user innovation, a great deal of innovation is done by those actually implementing and using technologies and products as part of their normal activities. In most of the times user innovators have some personal record motivating them. Sometimes user-innovators may become entrepreneurs, selling their product, they may choose to trade their innovation in exchange for other innovations, or they may be adopted by their suppliers. Nowadays, they may also choose to freely reveal their innovations, using methods like open source. In such networks of innovation the users or communities of users can further develop technologies and reinvent their social meaning.

Goals and Failures

Programs of organizational innovation are typically tightly linked to organizational goals and objectives, to the business plan, and to market competitive positioning. One driver for innovation programs in corporations is to achieve growth objectives. As Davila et al. (2006) notes, "Companies cannot grow through cost reduction and reengineering alone... Innovation is the key element in providing aggressive top-line growth, and for increasing bottom-line results".

One survey across a large number of manufacturing and services organizations found, ranked in decreasing order of popularity, that systematic programs of organizational innovation are most frequently driven by: Improved quality, Creation of new markets, Extension of the product range, Reduced labor costs, Improved production processes, Reduced materials, Reduced environmental damage, Replacement of products/services, Reduced energy consumption, Conformance to regulations.

These goals vary between improvements to products, processes and services and dispel a popular myth that innovation deals mainly with new product development. Most of the goals could apply to any organisation be it a manufacturing facility, marketing firm, hospital or local government. Whether innovation goals are successfully achieved or otherwise depends greatly on the environment prevailing in the firm.

Conversely, failure can develop in programs of innovations. The causes of failure have been widely researched and can vary considerably. Some causes will be external to the organization and outside its influence of control. Others will be internal and ultimately within the control of the organization. Internal causes of failure can be divided into causes associated with the cultural infrastructure and causes associated with the innovation process itself. Common causes of failure within the innovation process in most organizations can be distilled into five types: Poor goal definition, Poor alignment of actions to goals, Poor participation in teams, Poor monitoring of results, Poor communication and access to information.

Diffusion

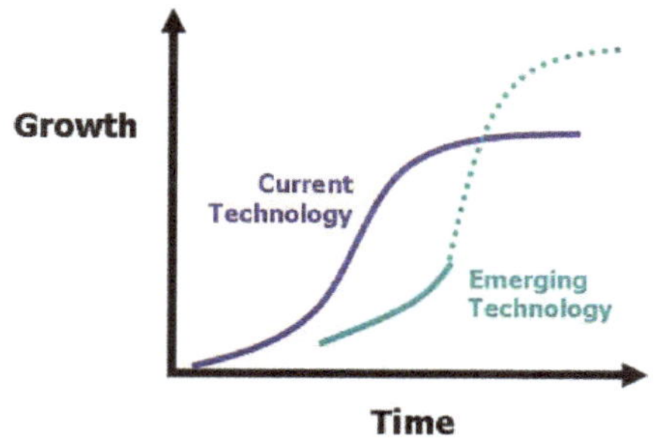

Diffusion of innovations

Diffusion of innovation research was first started in 1903 by seminal researcher Gabriel Tarde, who first plotted the S-shaped diffusion curve. Tarde defined the innovation-decision process as a series of steps that includes:

1. First knowledge
2. Forming an attitude
3. A decision to adopt or reject
4. Implementation and use
5. Confirmation of the decision

Once innovation occurs, innovations may be spread from the innovator to other individuals and groups. This process has been proposed that the life cycle of innovations can be described using the 's-curve' or diffusion curve. The s-curve maps growth of revenue or productivity against time. In the early stage of a particular innovation, growth is relatively slow as the new product establishes itself. At some point customers begin to demand and the product growth increases more rapidly. New incremental innovations or changes to the product allow growth to continue. Towards the end of its lifecycle, growth slows and may even begin to decline. In the later stages, no amount of new investment in that product will yield a normal rate of return

The s-curve derives from an assumption that new products are likely to have "product life" – i.e., a start-up phase, a rapid increase in revenue and eventual decline. In fact the great majority of innovations never get off the bottom of the curve, and never produce normal returns.

Innovative companies will typically be working on new innovations that will eventually replace older ones. Successive s-curves will come along to replace older ones and continue to drive growth upwards. In the figure above the first curve shows a current technology. The second shows an emerging technology that currently yields lower growth but will eventually overtake current technology and lead to even greater levels of growth. The length of life will depend on many factors.

Measures

Henry et al. in their review of literature on innovation management found 232 innovation metrics. They categorized these measures along five dimensions i.e. inputs to the innovation process, output from the innovation process, effect of the innovation output, measures to assess the activities in an innovation process and availability of factors that facilitate such a process.

There are two different types of measures for innovation: the organizational level and the political level.

Organizational Level

The measure of innovation at the organizational level relates to individuals, team-level assessments, and private companies from the smallest to the largest company. Measure of innovation for organizations can be conducted by surveys, workshops, consultants, or internal benchmarking. There is today no established general way to measure organizational innovation. Corporate measurements are generally structured around balanced scorecards which cover several aspects of innovation such as business measures related to finances, innovation process efficiency, employees' contribution and motivation, as well benefits for customers. Measured values will vary widely between businesses, covering for example new product revenue, spending in R&D, time to market, customer and employee perception & satisfaction, number of patents, additional sales resulting from past innovations.

Political Level

For the political level, measures of innovation are more focused on a country or region competitive advantage through innovation. In this context, organizational capabilities can be evaluated through various evaluation frameworks, such as those of the European Foundation for Quality Management. The OECD Oslo Manual (1995) suggests standard guidelines on measuring technological product and process innovation. Some people consider the Oslo Manual complementary to the Frascati Manual from 1963. The new Oslo manual from 2005 takes a wider perspective to innovation, and includes marketing and organizational innovation. These standards are used for example in the European Community Innovation Surveys.

Other ways of measuring innovation have traditionally been expenditure, for example, investment in R&D (Research and Development) as percentage of GNP (Gross National Product). Whether this is a good measurement of innovation has been widely discussed and the Oslo Manual has incorporated some of the critique against earlier methods of measuring. The traditional methods of measuring still inform many policy decisions. The EU Lisbon Strategy has set as a goal that their average expenditure on R&D should be 3% of GDP.

Indicators

Many scholars claim that there is a great bias towards the "science and technology mode" (S&T-mode or STI-mode), while the "learning by doing, using and interacting mode" (DUI-mode) is widely ignored. For an example, that means you can have the better high tech or software, but there are also crucial learning tasks important for innovation. But these measurements and research are rarely done.

A common industry view (unsupported by empirical evidence) is that comparative cost-effectiveness research (CER) is a form of price control which, by reducing returns to industry, limits R&D expenditure, stifles future innovation and compromises new products access to markets. Some academics claim the CER is a valuable value-based measure of innovation which accords truly significant advances in therapy (those that provide "health gain") higher prices than free market mechanisms. Such value-based pricing has been viewed as a means of indicating to industry the type of innovation that should be rewarded from the public purse. The Australian academic Thomas Alured Faunce has developed the case that national comparative cost-effectiveness assessment systems should be viewed as measuring "health innovation" as an evidence-based concept distinct from valuing innovation through the operation of competitive markets (a method which requires strong anti-trust laws to be effective) on the basis that both methods of assessing innovation in pharmaceuticals are mentioned in annex 2C.1 of the AUSFTA.

Rate of Innovation

Innovation Indices

Several indices attempt to measure innovation and rank entities based on these measures, such as:

- The Bloomberg Innovation Index
- The Bogota Manual, similar to the Oslo Manual, is focused on Latin America and the Caribbean countries.
- The Creative Class developed by Richard Florida
- The EIU Innovation Ranking
- The Global Competitiveness Report
- The Global Innovation Index (GII), by INSEAD

- The Information Technology and Innovation Foundation (ITIF) Index
- The Innovation Capacity Index (ICI) published by a large number of international professors working in a collaborative fashion. The top scorers of ICI 2009–2010 were: 1. Sweden 82.2; 2. Finland 77.8; and 3. United States 77.5.
- The Innovation Index, developed by the Indiana Business Research Center, to measure innovation capacity at the county or regional level in the United States.
- The Innovation Union Scoreboard
- The Innovations Indikator
- The INSEAD Innovation Efficacy Index (aka the INSEAD Innovation Efficiency Index)
- The International Innovation Index, produced jointly by The Boston Consulting Group (BCG), the National Association of Manufacturers (NAM), and The Manufacturing Institute (MI) (the NAM's nonpartisan research affiliate), is a worldwide index measuring the level of innovation in a country. NAM describes it as the "largest and most comprehensive global index of its kind".
- The Management Innovation Index - Model for Managing Intangibility of Organizational Creativity: Management Innovation Index
- The NYCEDC Innovation Index, by the New York City Economic Development Corporation, tracks New York City's "transformation into a center for high-tech innovation. It measures innovation in the City's growing science and technology industries and is designed to capture the effect of innovation on the City's economy."
- The Oslo Manual, similar to the Bogota Manual, is focused on North America, Europe, and other rich economies.
- The State Technology and Science Index, developed by the Milken Institute, is a U.S.-wide benchmark to measure the science and technology capabilities that furnish high paying jobs based around key components.
- The World Competitiveness Scoreboard

Innovation Rankings

Many research studies try to rank countries based on measures of innovation. Common areas of focus include: high-tech companies, manufacturing, patents, post secondary education, research and development, and research personnel. The following is a ranking of the top 10 countries in the 2015 Bloomberg Innovation Index:

Rank	Country/Territory
1	South Korea
2	Japan
3	Germany
4	Finland
5	Israel
6	United States
7	Sweden
8	Singapore
9	France
10	United Kingdom

Future of Innovation

Jonathan Huebner, a physicist working at the Pentagon's Naval Air Warfare Center, argued on the basis of both U.S. patents and world technological breakthroughs, per capita, that the rate of human technological innovation peaked in 1873 and has been slowing ever since. In his article, he asked "Will the level of technology reach a maximum and then decline as in the Dark Ages?" In later comments to *New Scientist* magazine, Huebner clarified that while he believed that we will reach a rate of innovation in 2024 equivalent to that of the Dark Ages, he was not predicting the reoccurrence of the Dark Ages themselves.

His paper received some mainstream news coverage at the time.

The claim has been met with criticism by John Smart, who asserted that research by technological singularity researcher Ray Kurzweil and others showed a "clear trend of acceleration, not deceleration" when it came to innovations. The foundation issued a reply to Huebner in the pages of the journal his article was published in, citing the existence of Second Life and eHarmony as proof of accelerating innovation; Huebner also replied to this. However, in 2010, Joseph A. Tainter, Deborah Strumsky, and José Lobo confirmed Huebner's findings using U.S. Patent Office data. Additional verification was provided in a 2012 paper by Robert J. Gordon.

Innovation and International Development

The theme of innovation as a tool to disrupting patterns of poverty has gained momentum since the mid-2000s among major international development actors such as DFID, Gates Foundation's use of the Grand Challenge funding model, and USAID's Global Development Lab. Networks have been established to support innovation in development, such as D-Lab at MIT. Investment funds have been established to identify and catalyze innovations in developing countries, such as DFID's Global Innovation Fund, Human Development Innovation Fund, and (in partnership with USAID) the Global Development Innovation Ventures.

Government Policies

Given the noticeable effects on efficiency, quality of life, and productive growth, innovation is a key factor in society and economy. Consequently, policymakers have long worked to develop environments that will foster innovation and its resulting positive benefits, from funding Research and Development to supporting regulatory change, funding the development of innovation clusters, and using public purchasing and standardisation to 'pull' innovation through.

For instance, experts are advocating that the U.S. federal government launch a National Infrastructure Foundation, a nimble, collaborative strategic intervention organization that will house innovations programs from fragmented silos under one entity, inform federal officials on innovation performance metrics, strengthen industry-university partnerships, and support innovation economic development initiatives, especially to strengthen regional clusters. Because clusters are the geographic incubators of innovative products and processes, a cluster development grant program would also be targeted for implementation. By focusing on innovating in such areas as precision manufacturing, information technology, and clean energy, other areas of national concern would be tackled including government debt, carbon footprint, and oil dependence. The U.S. Economic Development Administration understand this reality in their continued Regional Innovation Clusters initiative. In addition, federal grants in R&D, a crucial driver of innovation and productive growth, should be expanded to levels similar to Japan, Finland, South Korea, and Switzerland in order to stay globally competitive. Also, such grants should be better procured to metropolitan areas, the essential engines of the American economy.

Many countries recognize the importance of research and development as well as innovation including Japan's Ministry of Education, Culture, Sports, Science and Technology (MEXT); Germany's Federal Ministry of Education and Research; and the Ministry of Science and Technology in the People's Republic of China. Furthermore, Russia's innovation programme is the Medvedev modernisation programme which aims at creating a diversified economy based on high technology and innovation. Also, the Government of Western Australia has established a number of innovation incentives for government departments. Landgate was the first Western Australian government agency to establish its Innovation Program. The Cairns Region established the Tropical Innovation Awards in 2010 open to all businesses in Australia. The 2011 Awards were extended to include participants from all Tropical Zone Countries.

Innovators

An innovator, in a general sense, is a person or an organization, who is one of the first to introduce into reality, something better than before. Something that opens up a new area for others, and achieves an innovation.

Open Innovation

Open innovation is a term promoted by Henry Chesbrough, adjunct professor and faculty director of the Center for Open Innovation at the Haas School of Business at the University of California, in a book of the same name, though the idea and discussion about some consequences (especially the interfirm cooperation in R&D) date as far back as the 1960s. The term refers to the use of both inflows and outflows of knowledge to improve internal innovation and expand the markets for external exploitation of innovation.

The concept is also related to user innovation, cumulative innovation, know-how trading, mass innovation and distributed innovation.

"Open innovation is a paradigm that assumes that firms can and should use external ideas as well as internal ideas, and internal and external paths to market, as the firms look to advance their technology". Alternatively, it is "innovating with partners by sharing risk and sharing reward." The boundaries between a firm and its environment have become more permeable; innovations can easily transfer inward and outward.

The central idea behind open innovation is that, in a world of widely distributed knowledge, companies cannot afford to rely entirely on their own research, but should instead buy or license processes or inventions (i.e. patents) from other companies. In addition, internal inventions not being used in a firm's business should be taken outside the company (e.g. through licensing, joint ventures or spin-offs).

The open innovation paradigm can be interpreted to go beyond just using external sources of innovation such as customers, rival companies, and academic institutions, and can be as much a change in the use, management, and employment of intellectual property as it is in the technical and research driven generation of intellectual property. In this sense, it is understood as the systematic encouragement and exploration of a wide range of internal and external sources for innovative opportunities, the integration of this exploration with firm capabilities and resources, and the exploitation of these opportunities through multiple channels.

Advantages

Open innovation offers several benefits to companies operating on a program of global collaboration:

- Reduced cost of conducting research and development
- Potential for improvement in development productivity
- Incorporation of customers early in the development process
- Increase in accuracy for market research and customer targeting
- Potential for synergism between internal and external innovations
- Potential for viral marketing

Disadvantages

Implementing a model of open innovation is naturally associated with a number of risks and challenges, including:

- Possibility of revealing information not intended for sharing
- Potential for the hosting organization to lose their competitive advantage as a consequence of revealing intellectual property
- Increased complexity of controlling innovation and regulating how contributors affect a project
- Devising a means to properly identify and incorporate external innovation
- Realigning innovation strategies to extend beyond the firm in order to maximize the return from external innovation

Models of Open Innovation

Product Platforming

This approach involves developing and introducing a partially completed product, for the purpose of providing a framework or tool-kit for contributors to access, customize, and exploit. The goal is for the contributors to extend the platform product's functionality while increasing the overall value of the product for everyone involved.

Readily available software frameworks such as a software development kit (SDK), or an application programming interface (API) are common examples of product platforms. This approach is common in markets with strong network effects where demand for the product implementing the framework (such as a mobile phone, or an online application) increases with the number of developers that are attracted to use the platform tool-kit. The high scalability of platforming often results in an increased complexity of administration and quality assurance.

Idea Competitions

This model entails implementing a system that encourages competitiveness among contributors by rewarding successful submissions. Developer competitions such as hackathon events fall under this category of open innovation. This method provides organizations with inexpensive access to a large quantity of innovative ideas, while also providing a deeper insight into the needs of their customers and contributors.

Customer Immersion

While mostly oriented toward the end of the product development cycle, this technique involves extensive customer interaction through employees of the host organization. Companies are thus able to accurately incorporate customer input, while also allowing them to be more closely involved in the design process and product management cycle.

Collaborative Product Design and Development

Similarly to product platforming, an organization incorporates their contributors into the development of the product. This differs from platforming in the sense that, in addition to the provision of the framework on which contributors develop, the hosting organization still controls and maintains the eventual products developed in collaboration with their contributors. This method gives organizations more control by ensuring that the correct product is developed as fast as possible, while reducing the overall cost of development. Dr. Henry Chesbrough recently supported this model for open innovation in the optics and photonics industry.

Innovation Networks

Similarly to idea competitions, an organization leverages a network of contributors in the design process by offering a reward in the form of an incentive. The difference relates to the fact that the network of contributors are used to develop solutions to identified problems within the development process, as opposed to new products. Emphasis needs to be placed on assessing organisational capabilities to ensure value creation in open innovation.

Open Innovation in Science

In Austria the Ludwig Boltzmann Gesellschaft started a project named "Tell us!" about mental health issues and used the concept of open innovation to crowdsource research questions. The institute also launched the first "Lab for Open Innovation in Science" to teach 20 selected scientists the concept of open innovation over the course of one year. On Facebook the Ludwig Boltzmann Gesellschaft informs about the lab, the participants and teachers and on news on open innovation in science.

Closed Versus Open Innovation

The paradigm of closed innovation holds that successful innovation requires control. Particularly, a company should control the generation of their own ideas, as well as production, marketing, distribution, servicing, financing, and supporting. What drove this idea is that, in the early twentieth century, academic and government institutions were not involved in the commercial application of science. As a result, it was left up to other corporations to take the new product development cycle into their own hands. There just was not the time to wait for the scientific community to become more involved in the practical application of science. There also was not enough time to wait for other companies to start producing some of the components that were required in their final product. These companies became relatively self-sufficient, with little communication directed outwards to other companies or universities.

Throughout the years several factors emerged that paved the way for open innovation paradigms:

- The increasing availability and mobility of skilled workers
- The growth of the venture capital market
- External options for ideas sitting on the shelf
- The increasing capability of external suppliers

These four factors have resulted in a new market of knowledge. Knowledge is not anymore proprietary to the company. It resides in employees, suppliers, customers, competitors and universities. If companies do not use the knowledge they have inside, someone else will. Innovation can be generated either by means of closed innovation or by open innovation paradigms. There is an ongoing debate on which paradigm will dominate in the future.

Terminology of Open Innovation

Modern research of open innovation is divided into two groups, which have several names, but are similar in their essence (discovery and exploitation; outside-in and inside-out; inbound and outbound). The common factor for different names is the direction of innovation, whether from outside the company in, or from inside the company out:

Revealing (non-pecuniary outbound innovation)

This type of open innovation is when a company freely shares its resources with other partners, without an instant financial reward. The source of profit has an indirect nature and is manifested as a new type of business model.

Selling (pecuniary outbound innovation)

In this type of open innovation a company commercialises its inventions and technology through selling or licensing technology to a third party.

Sourcing (non-pecuniary inbound innovation)

This type of open innovation is when companies use freely available external knowledge, as a source of internal innovation. Before starting any internal R&D project a company should monitor the external environment in search for existing solutions, thus, in this case, internal R&D become tools to absorb external ideas for internal needs.

Acquiring (pecuniary inbound innovation)

In this type of open innovation a company is buying innovation from its partners through licensing, or other procedures, involving monetary reward for external knowledge

Open Source Versus Open Innovation

Open source and open innovation might conflict on patent issues. This conflict is particularly apparent when considering technologies that may save lives, or other open-source-appropriate technologies that may assist in poverty reduction or sustainable development. However, open source and open innovation are not mutually exclusive, because participating companies can donate their patents to an independent organization, put them in a common pool, or grant unlimited license use to anybody. Hence some open-source initiatives can merge these two concepts: this is the case for instance for IBM with its *Eclipse* platform, which the company presents as a case of open innovation, where competing companies are invited to cooperate inside an open-innovation network.

In 1997, Eric Raymond, writing about the open-source software movement, coined the term *the cathedral and the bazaar*. The cathedral represented the conventional method of employing a group of experts to design and develop software (though it could apply to any large-scale creative or innovative work). The bazaar represented the open-source approach. This idea has been amplified by a lot of people, notably Don Tapscott and Anthony D. Williams in their book Wikinomics. Eric Raymond himself is also quoted as saying that 'one cannot code from the ground up in bazaar style. One can test, debug, and improve in bazaar style, but it would be very hard to originate a project in bazaar mode'. In the same vein, Raymond is also quoted as saying 'The individual wizard is where successful bazaar projects generally start'.

Open-source specialist François Letellier advocates that open source (or free software) is a natural way of innovation in the software industry and that it is an exemplary and very effective form of open innovation, with open-source projects/communities act as innovation intermediaries.

Innovation System

The concept of the innovation system stresses that the flow of technology and information among people, enterprises and institutions is key to an innovative process. It contains the interaction between the actors who are needed in order to turn an idea into a process, product or service on the market.

Development and Diffusion of the Concept

Systems of Innovation are frameworks for understanding innovation which have become popular particularly among policy makers and innovation researchers first in Europe, but now anywhere in the world as in the 90's the World Bank and other UN affiliated institutions accepted. The concept of a 'system of innovation' was introduced by B.-Å. Lundvall in 1985 "however, as he and his colleagues would be the first to agree (and as Lundvall himself points out), the idea actually goes back at least to the Friedrich List´s conception of "The National System of Political Economy" (1841), which might just as well have been called "The National System of Innovation" (Freeman, 1995). Christopher Freeman coined the expression "National Innovation System" or in his 1988 study of the success of the Japanese economy. The concept, similarly used as "National System of Innovation" or "National Innovation System" was later applied to regions and sectors. According to innovation system theory, innovation and technology development are results of a complex set of relationships among actors in the system, which includes enterprises, universities and research institutes.

Innovation systems have been categorized into national innovation systems, regional innovation systems, local innovation systems, technological innovation systems and sectoral innovation systems.

There is no consensus on the exact definition of an innovation system, and the concept is still emerging. Innovation is often the result of the interaction among an ecology of actors, and the term 'innovation ecosystem' is occasionally used to emphasize this. For some, the expression 'innovation ecosystem' is a subset or synonym of 'innovation system'. Others separate between the expressions, using the expression "innovation system" for labeling a planned innovation environment, and "innovation ecosystem" for an ecological innovation environment.

Recently, the debate also started to study the problems that affect green innovation since in addition to the issues typical of innovation generally (such as market failures related to limited appropriability of economic benefits of knowledge), green growth innovation is also hindered by market failures related to the environment (pollution externalities). It is possible (and not uncommon) for an innovation system to successfully support innovation in many technology areas, but not in ones related to green growth. For this reason, it is necessary to focus on addressing both kinds of failures in order to drive innovation towards a green growth trajectory.

Examples of Definitions of National Innovation Systems

A national system of innovation has been defined as follows:

- " .. the network of institutions in the public and private sectors whose activities and interactions initiate, import, modify and diffuse new technologies." (Freeman, 1987)
- " .. the elements and relationships which interact in the production, diffusion and use of new, and economically useful, knowledge ... and are either located within or rooted inside the borders of a nation state." (Lundvall, 1992)
- "... a set of institutions whose interactions determine the innovative performance ... of national firms." (Nelson, 1993)
- " .. the national institutions, their incentive structures and their competencies, that determine the rate and direction of technological learning (or the volume and composition of change generating activities) in a country." (Patel and Pavitt, 1994)
- ".. that set of distinct institutions which jointly and individually contribute to the development and diffusion of new technologies and which provides the framework within which governments form and implement policies to influence the innovation process. As such it is a system of interconnected institutions to create, store and transfer the knowledge, skills and artefacts which define new technologies." (Metcalfe, 1995)
- ".. a human social network that behaves like a sociobiological system, wherein people have developed patterns of behavior that minimize transaction costs caused by social barriers resulting from geography, lack of trust, differences in language and culture, and inefficient social networks." (Hwang and Horowitt, 2012)

Social Innovation

Social innovations are new strategies, concepts, ideas and organizations that meet the social needs of different elements which can be from working conditions and education to community development and health — they extend and strengthen civil society. Social innovation includes the social *processes* of innovation, such as open source methods and techniques and also the innovations which have a social purpose — like online volunteering, microcredit, or distance learning.

Prominent innovators associated with the term include Pakistani Akhter Hameed Khan, Bangladeshi Muhammad Yunus, the founder of Grameen Bank which pioneered the concept of microcredit for supporting innovators in multiple developing countries in Asia, Africa and Latin America, and also inspired programs such as the Infolady Social Entrepreneurship Programme of Dnet (A Social Enterprise) and Stephen Goldsmith, former Indianapolis mayor who engaged the private sector in providing many city services.

Matters of Definition

Social Innovation has an inter-sectoral approach and is universally applicable. Social Innovations are launched by a variety of actors, including research institutions, companies or independent organizations, which each tend to use their own definition of Social Innovation. Therefore, it is worth discussing the most important aspects that distinguish it from other forms of social work or innovation. Social Innovation focuses on the process of innovation, how innovation and change take shape (as opposed to the more traditional definition of social innovation, giving priority to the internal organization of firms serving the productivity). Social Innovation focuses on new work and new forms of cooperation (business models)., especially those that work towards a sustainable society.

The Young Foundation, in order to distinguish between social and business innovation, stressed that social innovation is developed and diffused via organisations, whose primary purposes are not centred on mere profit maximisation (Mulgan et al., 2007, p. 8). The Bureau of European Policy Advisers more precisely defined social innovation as socially oriented in both ends and means (Hubert, 2010). According to these influential definitions, social innovation is characterised by: the capacity to address social needs that traditional policy seems increasingly unable to tackle; the empowerment of groups and individuals; and the willingness to change social relations. Hence, social innovation is often presented as a way to increase the quality of social services and their cost-effectiveness, offering equivalent, if not superior, outcomes despite considerable budget constraints.

Social innovation can take place within government; the for-profit sector, the nonprofit sector (also known as the third sector), or in the spaces between them. Research has focused on the types of platforms needed to facilitate such cross-sector collaborative social innovation.

Social entrepreneurship, like social enterprise, is typically in the nonprofit sector excluding both for-profit and public organizations. Both social entrepreneurship and social enterprise are important contributions to social innovation by creating social value and introducing new ways of achieving goals. Social entrepreneurship brings "new patterns and possibilities for innovation" and are willing to do things that existing organizations are not willing to do.

Social Innovation is often an effort of mental creativity which involves fluency and flexibility from a wide range of disciplines. The act of social innovation in a sector is mostly connected with diverse disciplines within the society. The social innovation theory of 'connected difference' emphasizes three key dimensions to social innovation. First, innovations are usually new combinations or hybrids of existing elements, rather than completely new. Second, their practice involves cutting across organizational or disciplinary boundaries. Lastly, they leave behind compelling new relationships between previously separate individuals and groups. Social innovation is also gaining visibility within academia.

History

Social innovation was discussed in the writings of figures such as Peter Drucker and Michael Young (founder of the Open University and dozens of other organizations) in the 1960s. It also appeared in the work of French writers in the 1970s, such as Pierre Rosanvallon, Jacques Fournier, and Jacques Attali. However, the themes and concepts in social innovation existed long before. Benjamin Franklin, for example, talked about small modifications within the social organization of communities that could help to solve everyday problems. Many radical 19th century reformers like Robert Owen, founder of the cooperative movement, promoted innovation in the social field and all of the great sociologists including Karl Marx, Max Weber and Émile Durkheim focused attention on broader processes of social change. Other theories of innovation became prominent in the 20th century, many of which had social implications, without putting social progress at the center of the theory. Joseph Schumpeter, for example, addressed the process of innovation directly with his theory of creative destruction and his definition of entrepreneurs as people who combined existing elements in new ways to create a new product or service. Beginning in the 1980s, writers on technological change increasingly addressed how social factors affect technology diffusion.

The article "Rediscovering Social Innovation" mentions how social innovations are dependent on history and the change in institutions. The article discusses the ten recent social innovations reflecting current change to include:

- Charter Schools: Charter schools are a social innovation that provides an alternative avenue for students to continue to develop and build upon their educational foundation without many of the issues prominent in the public school system. These primary and secondary schools are publicly funded and operate independently, which allows the teachers and parents to collaboratively develop alternative teaching methods for their students as related regulations are less stringent for Charter Schools.

- Community-Centered Planning: This social innovation allows communities to plan and develop systems that cater solutions to their specific local needs by using their historical knowledge and other local resources. .

- Emissions Trading: The Emissions Trading program was designed to address issues associated with the continuous increase in pollution. The program provides solutions such as setting a cap on the amount that certain pollutants can be emitted, and implementing a permit system to control the amount of pollution produced by each participating business. If a business needs to use more pollution than permitted, it can purchase credits from a business that has not emitted its maximum permitted amount. The goal of the Emissions Trading program is that, over time and with increased awareness, society will limit the types and the numbers of pollutants emitted to what is only necessary.

- Fair Trade: Products including coffee, sugar, and chocolate are currently being traded without high standards that result in tough conditions for farmers and a less sustainable environment. Fair trade is a movement that certifies traders to exchange with the farmers that produce these products. The idea behind this movement is that by being paid a living-wage, being able to meet social and environmental standards and promoting "environmental sustainability, the lives of these farmers will be improved.
- Habitat Conservation Plans: Habitat Conservation Plans is an effort by the US Fish and Wild Life Service and the Environmental Protection Agency to protect species and their endangerment by providing economical incentives to conserve their habitats and protect these species from endangerment.
- Individual Development Accounts: This social innovation is made to support the working poor with saving decisions that they have made to better enhance their lives. This initiative will give $2 per every $1 saved by the working poor for College tuition, purchasing a home, starting a business, and other similar and productive initiatives. This is made possible by philanthropic, government and corporate sponsors that donate to this cause.
- International Labor Standards: Labor standards differ country-to-country, with some agreeably better than others. In effort to internationally align these, the International Labor Organization, participating governments, and employees contributed to the development of standards that protect workers' rights to freedom, equity, security, and human dignity".
- Microfinance: This social innovation is created to support the those financially unable to gain access to financial services such as banking, lending, and insurance. The ultimate goal of Microfinance is to enable an escape from poverty by helping to improve the living conditions and financial viability among the impoverished program participants.
- Socially Responsible Investing: "An investment strategy that attempts to maximize both financial and social returns. Investors generally favor businesses and other organizations whose practices support environmental sustainability, human rights, and consumer protection."
- Supported Employment: Supported Employment is a social innovation geared towards helping disabled or disadvantaged workers who are un- or under-employed due to their condition obtain suitable employment. The Support Em-

ployment service provides access to job coaches, transportation, assistive technology, specialized job training, and individual tailored supervision in effort to help program participants become more competitive applicants and better prepared overall for the job market.

Criticism

Over the last two decades social innovation has gained significant popularity as a strategy to tackle new social risks including population ageing and its health correlates (Hubert, 2010; Mulgan et al., 2007, 2010; Murray, Caulier-Grice and Mulgan, 2010). However, as other concepts recently developed within the academic debate – among them, social capital (Ferragina, 2012) – social innovation might soon turn out to be simply another way to juxtapose the qualifier "social" to the private sector jargon, in order to avoid heated discussions on structural inequalities (Grisolia and Ferragina, 2015).

In the context of 'neoliberal austerity', a strong call in favour of social innovation might hide the attempt to shift public attention from structural deficiencies and disparities to individual and group responsibility, following the vision: "doing more with less" . In order to guarantee universal coverage and universal social rights, however, the welfare state system cannot be managed with the logic of mere cost-effectiveness alone (Grisolia and Ferragina, 2015). A Universal coverage is the precondition for any well-functioning economy, not the other way around. As such, the enhancement of "politically motivated policies under the pretence of budget cuts" can be particularly dangerous in its consequences for population health (Kleinert and Horton, 2013, p. 1074). Social innovation per se might not be able to substantially tackle pressing social needs. Rather, the all-innovating and self-empowering jargon currently in vogue might disguise a dangerous inattention to structural inequalities, adversely affecting health outcomes across the board, but especially of the poorest. Among the therapies prescribed by the neoliberal orthodoxy – liberalisation, deregulation, devolution, individual or group empowerment – social innovation might soon reveal itself as a convenient buzzword, an eclectic concept to dissimulate political choices, legitimated by the doctrine of budgetary constraints. The redistribution of resources "from past to present generations" – keeping constant the overall public spending – and the shift from a "transfer-based" to a "service-based" welfare state would represent a truly innovative approach to social policy, offering a credible and responsible alternative to the magic wand of social innovation.

Developments Since 2000

Academic research, blogs and websites feature social innovation, along with organizations working on the boundaries of research and practical action. Topics include:

- Innovation in public services was pioneered particularly in some Scandinavian and Asian countries. Governments are increasingly recognizing that innovation requires healthcare, schooling and democracy.

- Social entrepreneurship, which is the practice of creating new organizations focusing on non-market activities.
- Responsible Research and Innovation, which takes into account effects and potential impacts on the environment and society. It includes Engagement of all societal actors (researchers, industry, policymakers and civil society); Gender Equality; Science Education; Open Access; Ethics; and Governance.
- Online volunteering, a free service launched in 2000 whereby individuals from all over the world contribute to the needs of development organizations and public institutions
- Open source innovation, in which the intellectual property involved in a product or service is made freely available.
- Complex adaptive systems, which have built-in mechanisms to help them adapt to changing circumstances.
- Collaborative approaches which involve stakeholders who are not directly responsible for some activity, such as stockholders and unions collaborating on business issue and business collaborating with government on regulatory issues.
- Innovation diffusion
- Localized influences that make some localities particularly innovative.
- Institutional or system entrepreneurship which focuses on agents who work at a broad system level in order to create the conditions which will allow innovations to have a lasting impact.
- Business, particularly in services.

Institutional Support

The US created an Office for Social Innovation in the White House, which is funding projects that combine public and private resources. with foundations that support social innovation. In 2010, the US government listed 11 investments made by its 'Social Innovation Fund', with public funding more than matched by philanthropic organizations. This fund focuses on partnerships with charities, social enterprises, and business. Moreover, educational institutions are now increasingly supporting teaching and research in the area of social innovation. In addition to pioneered efforts by institutions such as the Harvard Business School's Initiative on Social Enterprise (launched 1993) and Said Business School's Skoll Centre for Social Entrepreneurship (launched 2003), INSEAD and other universities now offer short-term programs in Social Innovation, and a few such as Goldsmiths, University of London offer Masters courses dedicated entirely to the study of theory and practice in relation to social entrepreneurship and innovation.

Public policy makers support social innovation in the UK, Australia, China and Denmark, as well. The European Union's innovation strategy was the first well-funded research and development strategy to emphasize social innovation.

In 2002, the South Australian government, led by Premier and Social Inclusion Minister Mike Rann, embraced a ten-year social innovation strategy with big investments and a focus on reform in areas such as homelessness, school retention, mental health and disability services.

The Common Ground and Street to Home homelessness initiatives and the Australian Centre for Social Innovation were established in Adelaide and many reforms trialed in South Australia have been adopted nationally throughout Australia. This initiative, headed by Monsignor David Cappo, South Australia's Social Inclusion Commissioner, was advised by 'Thinkers in Residence' Geoff Mulgan and New York social entrepreneur Rosanne Haggerty.

Local and Regional development

Literature on social innovation in relation to territorial/regional development covers innovation in the social economy, i.e. strategies for satisfaction of human needs; and innovation in the sense of transforming and/or sustaining social relations, especially governance relations at the regional and local level. Beginning in the late 1980s, Jean-Louis Laville and Frank Moulaert researched social innovation. In Canada CRISES initiated this type of research. Another, larger project was SINGOCOM a European Commission Framework 5 project, which pioneered so-called "Alternative Models for Local Innovation" (ALMOLIN). These models were further elaborated through community actions covered by KATARSIS and SOCIAL POLIS. More recent works focus on the societal role of the economic life in terms of innovations in social practices and social relations at the local and regional levels. Social Innovation, therefore, is increasingly seen as a process and a strategy to foster human development through solidarity, cooperation, and cultural diversity.

The EU funded URBACT programme is designed to help cities to exchange and learn around urban policies. The URBACT methodology can be seen as a social innovation action planning approach. A typical URBACT network would have ten cities working on a specific theme such as active inclusion or regenerating disadvantaged neighbourhoods. They examine good practice and then working through a local support group use the results to inform their local action plan.

The Social Innovation Europe initiative, funded by the European Commission's Directorate General for Enterprise and Industry, was set up to map social innovation at a European level, by creating a directory of grass-roots examples of social innovation from across the 27 member states.

The European Commission funded the SELUSI study between 2008 - 2013 that looked at over 550 social ventures and examined how these insights can spark change and innovation at a much larger scale. It looked at business models of social ventures in five countries - UK being one of them – identifying which specific practices evolved by social ventures are particularly successful, and how and by whom – be it social enterprise, public sector body or mainstream business– they can be most effectively scaled-up.

The European Commission has launched a new initiative (project) in 2013 under FP7 funding – BENISI Project – with the aim to build a network of incubators for social innovation across regions and countries. This network facilitates identification of 300 social innovation examples and facilitates its scaling. The network is organised in a way to identify new models for scaling of social innovations across various geographical clusters in collaboration with each other, communicating the ideas, finding the tools and funds, developing business plans and models in order to promote the new promising ideas throughout Europe.

- A guide also exists that provides a way to promote social innovations at a local or regional level.

Innovation Intermediary

Innovation Intermediaries is a concept in innovation studies to help understand the role of firms, agencies and individuals that facilitate innovation by providing the bridging, brokering, knowledge transfer necessary to bring together the range of different organisations and knowledge needed to create successful innovation. Innovation intermediaries have always played a key role in innovation, for example the agricultural middlemen in 16th, 17th and 18th century Britain who not only bought and sold wool, but facilitated the transfer of knowledge of new techniques. With the recent fashion for 'Open Innovation' involving complex networks of firms and users, organisations such as consultants, service incubators, conference organisers, trade organisations, government innovation agencies etc. are now recognised as playing a central role in facilitating and coordinating innovation.

Innovation intermediaries are variously described as 'bridgers', ' change agents', 'brokers'. They are important as the developers of a new invention or technique are seldom connected to their potential users, or to the firms and organisations that have complementary expertise, knowledge and resources. The same applies to potential users of innovations, so that intermediaries are needed to bring organisations and knowledge together to build supply networks and markets.

Intermediaries have also been defined as a system of complementary organizational categories that shape, pilot and ensure systemic integration, by reducing the complexity of transactions, enabling institutional change and promoting crucial learning dynamics among system components, organizations and entrepreneurs; across political, economic and social innovation-relevant levels. These categories could be settled together, permitting a holistic approximation to the matter of intermediation. The novel notion of a system of intermediary organizations could also facilitate the coordination and evaluation of their profiles and missions over time and space-based requirements.

Intermediaries play a wide range of roles, facilitating the bringing together of various actors at different parts of innovation processes such as ideation, invention, standards making, managing IPR, commercialisation, creating new market segments etc. These intermediaries can specialise in different services. Basic functions include process coordination and matchmaking between innovation seekers and potential solution providers, knowledge and finance broking, testing, standardisation, project valuation and portfolio management etc. Each of these activities facilitates the exchange and the building of new knowledge, creates opportunities for experimentation, helps the emergence of standards and common goals, and the formation of partnerships.

From an economic perspective, innovation intermediaries create multi-sided markets by delivering value to several agents on a matching market, create the market and manage the matching process.

In the age of the internet, there are increasing numbers of online services and platform that explicitly try to play the role of innovation intermediary, including technology matchmaking services, crowdfunding and crowdsourcing services, meetup sites etc. Despite the trend of online platforms, a managerial role for effective tie building in networks or efficient technology matchmaking is necessary.

Pace of Innovation

Pace of innovation is the speed at which technological innovation or advancement is occurring, with the most apparent instances being too slow or too rapid. Both these rates of innovation are extreme and therefore have effects on the people that get to use this technology. The factors that cause rapid pace of innovations include lack of trial and error, technological momentum, ease of entry into a field. Innovators usually do not have the burden of proof, thus regulatory agencies have to prove that the technology has negative effects. Slow pace of innovation is caused by, but not limited to, lack of monetary funding, conflict of interests between the innovators and other factions such as the consumers, and technological momentum of competing technologies which make the use of new innovations difficult. Different technologies require different paces of innovation, therefore it is important to note that the appropriate pace of innovation depends on the context.

Technology Moving too Slowly

A technology with a slow pace is one that has its developments delayed or under-emphasized due to different factors or barriers. Major barriers include technosocial momentum that advance an alternative innovation, lack of adequate monetary funds, and social factors. Non-automobile based cities progress too slowly today in the USA in comparison to those that are based around automobiles. This acts as a barrier to high speed trains and other forms of mass transit in most cities. Another example of a slowed technology is the male birth control pill which has been researched since the 1920s and still a working product has not ended up on the market. The slow progress and acceptance of the male birth control pill is largely associated with budget issues where it lacks the funding to perform proper research. A male contraceptive that is nearly 100% effective has yet to be found, and testing can run millions of dollars. Conversely, since many of the technologies are long-term solutions for men, pharmaceutical companies are resistant to further development of products that are not constantly making money. Those contraceptives that have been found to work, whether they are pills or herbal remedies, but they have adverse side effects that are less than desired by men. This social barrier has also slowed the progress of the male birth control pill as men do not naturally experience the emotional feelings of menstruation and pregnancy.

The slow pace of technology not only affects products, but entire fields, such as green chemistry. Social and organizational barriers have prevented green chemistry from becoming a viable field, whereas brown chemistry's technological momentum remains strong. A lack of environmental considerations in professional licensing exams for chemical engineers prevents a shift to a more benign chemical education that reflects the integrity of the natural planet. Since this field is relatively new, definitions and standards are still vague, leading to organizational barriers. There is no large-scale green chemistry database or comprehensive reference, which limits the progress being made in the field. Companies that make green technologies or produce tools to measure green performance are often proprietary and not easily applicable to other firms. Barriers like these can cause a technology to stumble or to not move at all. This causes some goods not to go into production due to lack of funding. Examples of some products that have not been fully developed due to poor funding include the male birth control pill and the dengue fever vaccine.

Technology Moving too Fast

The description of technology moving too fast is that innovation is proceeding faster than expected or than the consumption rate of the goods due to a quick rate of production. The main reason why innovating too fast is undesirable is that the technology tends to have unintended consequences caused by the lack of adequate trial and error, including deliberation among all parties involved in the use of a product. There are numerous examples in history that point out the unintended consequences of innovating too fast. The manufacturing of the Ford Pinto during the 1960s through 1970s, occurred due to fast manufacturing of the vehicles before adequate testing could be done.

This led to the production of an ineffective product that caused the deaths and injuries of numerous people when the Ford Pinto was involved in an accident. Another example of innovating too fast is suburbanization, which occurred due to the ease of access into big cities in combination with lower suburban housing rates. This had the unintended consequence of reducing the tax funds for inner urban areas and undermined the effectiveness of public transport due to the long distances needed to be traveled. This also led to the increase of small vehicle sales, which played a part in the deterioration of the environment particularly due to air pollution. To create a market for the numerous goods produced, manufacturers create low quality goods that do not last for long and advertise extensively which creates an endless cycle of consumerism.

Lack of deliberation amongst the parties to be involved in a technology may also lead to a rapid pace of innovation mainly due to publicity that a project may gain making it difficult to oppose the original plan leading to the technological momentum in that particular field. For example, the NASA project on space shuttle lead to large publicity of the project which made it more difficult to stop the project when it stated being costly. Lack of deliberation during the decision made to build large nuclear reactors during the 1970s, caused nuclear energy to have a technological momentum. In the case of nuclear power plants, there was also no gradual scale up of the power plants which made it difficult and costly to implement changes after trial and error. Other examples of technology proceeding too quickly are product iterations that occur from the treadmill of consumption. With consumer goods, such as CD players, vacuums, and lint rollers, being produced with such variety that lead to various resource inefficiencies, products are being modified at an unnecessarily rapid pace.

Vertical Innovation

Vertical innovation is an innovation on the quality of the goods, it consist to improve the quality of items by innovation permitting to those items to access to the highest quality available on the economy. In the literature of growth it is oppose to "horizontal innovation" which in turn refer to product innovation mean the introduction of new items in the economy. 'vertical integration' and 'open innovation'. The term vertical integration describes a style of business management control, and is typified by one firm engaged in different multiple parts of a production process. Open innovation is a term promoted by Henry Chesbrough, a professor and executive director at the Center for Open Innovation at the University of California, Berkeley. "Open innovation is a paradigm that assumes that firms can and should use external ideas as well as internal ideas, and internal and external paths to market, as the firms look to advance their technology."

The company Nottingham Spirk coined the phrase 'vertical innovation' in 2011, to describe the proprietary process developed over the forty years at the firm. The process is a development strategy to take new product and service concepts and turn them into tangible items within a compressed timeframe. Nottingham Spirk serves clients as an open innovation partner and uses a vertically integrated model to execute on development; hence a 'vertical innovation' strategy.

Vertical innovation is both a process to encourage efficient problem solving and a corporate culture, where all development activities are handled by a cohesive and experienced team of experts. This group leverages the skills and talents of many individuals to move product development quickly forward. This process focuses on continuous improvement through each phase of a development program.

Different Product Development and Innovation Approaches

Vertical Innovation Approach

Vertical innovation ideally takes place at one location, providing many points of collaboration, continuity in communication and helps ensure confidentiality. Co-located skill groups have individuals participating in each of the typical development stages: Market Analysis, Customer Insights, Ideation and Design, Engineering and Prototyping, Packaging and Retail Experience, Sourcing and Production Support. This allows for continual improvement and optimization at many points along the development process, while providing shortened development timelines. The vertical innovation approach also encourages random connections between different skill sets to provide the best chances for new and innovative advancements.

A real-life example of a physical space that encourages creative collaboration is the building that houses Pixar, the computer animation studio that created innovative, Academy-award-winning blockbuster films like Toy Story, Monsters and Finding Nemo. As Walter Isaacson writes in his biography of Steve Jobs, Jobs designed the Pixar building to promote chance encounters and unplanned collaborations. "If a building doesn't encourage that, you'll lose a lot of innovation and magic that's sparked by serendipity," Jobs said. "So we designed the building to make people get out of their offices and mingle in the central atrium with people they might not otherwise see." The front doors and main stairs and corridors all lead to a central atrium, where a cafe and employee mailboxes are located as well. John Lasseter, Chief Creative Officer at Pixar, confirmed the success of the building's layout: 'Steve's theory worked from day one. I kept running into people I hadn't seen for months. I've never seen a building that promoted collaboration and creativity as well as this one.'

—A.Meyer, openinnovation.org

Common Innovation Approach

Corporations commonly rely on different companies, groups or separate individuals for different skill sets. They use separate internal groups or external organizations to provide the skills necessary for innovation development. For example:

- a research group from an outside firm gathers customer insights
- an internal design group creates concepts
- another firm develops engineering layouts
- a third company handles intellectual property review
- a fourth builds the prototypes
- a fifth firm creates packaging
- and a final group organizes supply chain management

This chain of hand-offs typically lengthens development timelines, hinders innovative ideas and dilutes responsibilities needed for project success.

Open Innovation Approach

The central idea behind open innovation is that in a world of widely distributed knowledge, companies cannot afford to rely entirely on their own research, but should instead buy or license skills, processes or inventions (i.e. patents) from other individuals or companies.

This format makes use of many different inputs of skills and data into a development program. Because these connections can occur from the inside or outside of a company, there is still a high need for analysis, coordination, testing and maintaining of confidentiality.

Open Innovation Intermediary

Open innovation intermediaries as described by Henry Chesbrough in his 2006 book *Open Business Models* are "companies that help other companies implement various facets of open innovation". The presence of these firms enables other companies to explore the market for ideas without getting in over their heads, since the intermediaries can act as guides to help those other companies along the trail. There are different kinds of open innovation intermediaries, with some acting as agents (representing one side of a transaction) while others take on the role of brokers or market-makers.

Fundamentals of Open Innovation Intermediaries

Open Innovation Intermediaries are companies that act as a mediators and incubators between other companies that have commenced in an Open Innovation activity. Henry Chesbrough himself is a chairman at the Open Innovation Center – Brazil, a company that fully operates as an Open Innovation Intermediary between two or more companies that need to embark into Open Innovation activities. The focus of the intermediaries is that companies should make much greater use of external ideas and technologies in their own business and in turn allow unused internal ideas to go outside for others to use in their business. There are three main areas that branch out from what Open Innovation Intermediaries focus on. One is business model innovation [Chesbrough. H., 2006 Open Business Models: How to Thrive in the New Innovation Landscape]. The second is management of intellectual property in a world of open innovation. And the third is innovation of services.

The Role of Open Innovation Intermediaries

Open Innovation Intermediaries are responsible for facilitating the Open Innovation activity that companies are undertaking, focusing on fully exploiting the benefits of the mutual action and thoroughly mitigating the disadvantages and risks for all of the companies.

Unbundling

Unbundling is a neologism to describe how the ubiquity of mobile devices, Internet connectivity, consumer web technologies, social media and information access in the 21st century is affecting older institutions (education, broadcasting, newspapers, games, shopping, etc.) by "break[ing] up the packages they once offered, providing particular parts of them at a scale and cost unmatchable by the old order." Unbundling has been called "the great disruptor".

Etymology

"Unbundling" most basically means simply the "process of breaking apart something into smaller parts." In the context of mergers and acquisitions, unbundling refers to the "process of taking over a large company with several different lines of business, and then, while retaining the core business, selling off the subsidiaries to help fund the takeover."

Examples

- Massive open online courses are "part of a trend towards the unbundling of higher education" by providing access to recorded lectures, online tests, and digital documents as a complement to traditional classroom instruction.
- Pandora Radio
- The addition of Maryland and Rutgers to NCAA football was described as part of a larger trend towards the unbundling of each university's broadcast rights to maintain profitability.

- The CEO of Mashable predicted that unbundled news contents' "microcontent sharing" via software like Flipboard (Android and iOS), Zite and Spun (iPhone) would be a major trend in 2013.
- LinkedIn has embraced a multi-app strategy and now has a family of six separate apps—The LinkedIn 'Mothership' app and 'satellite' apps ranging from job search to tailored news
- The customers that live in large apartment complexes and multiple dwelling units can be unbundled in a way that allows multiple providers to reach each of the different units.

References

- MacCallum, D., Moulaert, F., Hillier, J. and S. Vicari (Eds) (2009) Social Innovation and Territorial Development. Ashgate, Aldershot. ISBN 978-0-7546-7233-3
- Chesbrough, Henry William (1 March 2003). Open Innovation: The new imperative for creating and profiting from technology. Boston: Harvard Business School Press. ISBN 978-1578518371.
- Hwang, Victor and Greg Horowitt (2012). The Rainforest: The Secret to Building the Next Silicon Valley. Los Altos Hills: Regenwald. p. 304. ISBN 978-0615586724.
- Goldsmith, Stephen (March 2010). The Power of Social Innovation: How Civic Entrepreneurs Ignite Community Networks for Good. Jossey-Bass. ISBN 978-0-470-57684-7.
- "Bill & Melinda Gates Foundation and Grand Challenge Partners Commit to Innovation with New Investments in Breakthrough Science - Bill & Melinda Gates Foundation". Gatesfoundation.org. 2014-10-07. Retrieved 2016-03-14.
- "Global Development Lab | U.S. Agency for International Development". Usaid.gov. 2015-08-05. Retrieved 2016-03-14.
- "USAID and DFID Announce Global Development Innovation Ventures to Invest in Breakthrough Solutions to World Poverty | U.S. Agency for International Development". Usaid.gov. 2013-06-06. Retrieved 2016-03-14.
- ""Lab for Open Innovation"-Lehrgang der Ludwig Boltzmann Gesellschaft | PROFIL.at". profil.at. 2016-03-17. Retrieved 2016-06-19.

3

Essential Concepts of Innovation Management

The essential concepts of innovation management include confirmation bias, ambidextrous organization, disruptive innovation, leapfrogging, paradigm shift, eureka effect etc. The ability of an organization to be efficient is known as organizational ambidexterity whereas the tendency of an innovation to disrupt an existing market is known as disruptive innovation. The chapter strategically encompasses and incorporates the key concepts of innovation management, providing a complete understanding.

Confirmation Bias

Confirmation bias, also called confirmatory bias or myside bias, is the tendency to search for, interpret, favor, and recall information in a way that confirms one's preexisting beliefs or hypotheses, while giving disproportionately less consideration to alternative possibilities. It is a type of cognitive bias and a systematic error of inductive reasoning. People display this bias when they gather or remember information selectively, or when they interpret it in a biased way. The effect is stronger for emotionally charged issues and for deeply entrenched beliefs. People also tend to interpret ambiguous evidence as supporting their existing position. Biased search, interpretation and memory have been invoked to explain attitude polarization (when a disagreement becomes more extreme even though the different parties are exposed to the same evidence), belief perseverance (when beliefs persist after the evidence for them is shown to be false), the irrational primacy effect (a greater reliance on information encountered early in a series) and illusory correlation (when people falsely perceive an association between two events or situations).

A series of experiments in the 1960s suggested that people are biased toward confirming their existing beliefs. Later work re-interpreted these results as a tendency to test ideas in a one-sided way, focusing on one possibility and ignoring alternatives. In certain situations, this tendency can bias people's conclusions. Explanations for the observed biases include wishful thinking and the limited human capacity to process information. Another explanation is that people show confirmation bias because they are weighing up the costs of being wrong, rather than investigating in a neutral, scientific way.

Confirmation biases contribute to overconfidence in personal beliefs and can maintain or strengthen beliefs in the face of contrary evidence. Poor decisions due to these biases have been found in political and organizational contexts.

Types

Confirmation biases are effects in information processing. They differ from what is sometimes called the *behavioral confirmation effect*, commonly known as *self-fulfilling prophecy*, in which a person's expectations influence their own behavior, bringing about the expected result.

Some psychologists restrict the term *confirmation bias* to selective collection of evidence that supports what one already believes while ignoring or rejecting evidence that supports a different conclusion. Other psychologists apply the term more broadly to the tendency to preserve one's existing beliefs when searching for evidence, interpreting it, or recalling it from memory.

Biased Search for Information

Confirmation bias has been described as an internal "yes man", echoing back a person's beliefs like Charles Dickens' character Uriah Heep.

Experiments have found repeatedly that people tend to test hypotheses in a one-sided way, by searching for evidence consistent with their current hypothesis. Rather than searching through all the relevant evidence, they phrase questions to receive an affirmative answer that supports their theory. They look for the consequences that they would expect if their hypothesis were true, rather than what would happen if it were false. For example, someone using yes/no questions to find a number he or she suspects to be the number 3 might ask, "Is it an odd number?" People prefer this type of question, called a "positive test", even when a negative test such as "Is it an even number?" would yield exactly the same information. However, this does not mean that people seek tests that guarantee a positive answer. In studies where subjects could select either such pseudo-tests or genuinely diagnostic ones, they favored the genuinely diagnostic.

The preference for positive tests in itself is not a bias, since positive tests can be highly informative. However, in combination with other effects, this strategy can confirm existing beliefs or assumptions, independently of whether they are true. In real-world situations, evidence is often complex and mixed. For example, various contradictory ideas about someone could each be supported by concentrating on one aspect of his or her behavior. Thus any search for evidence in favor of a hypothesis is likely to succeed. One illustration of this is the way the phrasing of a question can significantly change the answer. For example, people who are asked, "Are you happy with your social life?" report greater satisfaction than those asked, "Are you *un*happy with your social life?"

Even a small change in a question's wording can affect how people search through available information, and hence the conclusions they reach. This was shown using a fictional child custody case. Participants read that Parent A was moderately suitable to be the guardian in multiple ways. Parent B had a mix of salient positive and negative qualities: a close relationship with the child but a job that would take him or her away for long periods of time. When asked, "Which parent should have custody of the child?" the majority of participants chose Parent B, looking mainly for positive attributes. However, when asked, "Which parent should be denied custody of the child?" they looked for negative attributes and the majority answered that Parent B should be denied custody, implying that Parent A should have custody.

Similar studies have demonstrated how people engage in a biased search for information, but also that this phenomenon may be limited by a preference for genuine diagnostic tests. In an initial experiment, participants rated another person on the introversion–extroversion personality dimension on the basis of an interview. They chose the interview questions from a given list. When the interviewee was introduced as an introvert, the participants chose questions that presumed introversion, such as, "What do you find unpleasant about noisy parties?" When the interviewee was described as extroverted, almost all the questions presumed extroversion, such as, "What would you do to liven up a dull party?" These loaded questions gave the interviewees little or no opportunity to falsify the hypothesis about them. A later version of the experiment gave the participants less presumptive questions to choose from, such as, "Do you shy away from social interactions?" Participants preferred to ask these more diagnostic questions, showing only a weak bias towards positive tests. This pattern, of a main preference for diagnostic tests and a weaker preference for positive tests, has been replicated in other studies.

Personality traits influence and interact with biased search processes. Individuals vary in their abilities to defend their attitudes from external attacks in relation to selective exposure. Selective exposure occurs when individuals search for information that is consistent, rather than inconsistent, with their personal beliefs. An experiment examined the extent to which individuals could refute arguments that contradicted their personal beliefs. People with high confidence levels more readily seek out contradictory information to their personal position to form an argument. Individuals with low confidence levels do not seek out contradictory information and prefer information that supports their personal position. People generate and evaluate evidence in arguments that are biased towards their own beliefs and opinions. Heightened confidence levels decrease preference for information that supports individuals' personal beliefs.

Another experiment gave participants a complex rule-discovery task that involved moving objects simulated by a computer. Objects on the computer screen followed specific laws, which the participants had to figure out. So, participants could "fire" objects across the screen to test their hypotheses. Despite making many attempts over a ten-hour session, none of the participants figured out the rules of the system. They typically attempted to confirm rather than falsify their hypotheses, and were reluctant to consider alternatives. Even after seeing objective evidence that refuted their working hypotheses, they frequently continued doing the same tests. Some of the participants were taught proper hypothesis-testing, but these instructions had almost no effect.

Biased Interpretation

Smart people believe weird things because they are skilled at defending beliefs they arrived at for non-smart reasons.

—Michael Shermer

Confirmation biases are not limited to the collection of evidence. Even if two individuals have the same information, the way they interpret it can be biased.

A team at Stanford University conducted an experiment involving participants who felt strongly about capital punishment, with half in favor and half against it. Each participant read descriptions of two studies: a comparison of U.S. states with and without the death penalty, and a comparison of murder rates in a state before and after the introduction of the death penalty. After reading a quick description of each study, the participants were asked whether their opinions had changed. Then, they read a more detailed account of each study's procedure and had to rate whether the research was well-conducted and convincing. In fact, the studies were fictional. Half the participants were told that one kind of study supported the deterrent effect and the other undermined it, while for other participants the conclusions were swapped.

The participants, whether supporters or opponents, reported shifting their attitudes slightly in the direction of the first study they read. Once they read the more detailed descriptions of the two studies, they almost all returned to their original belief regardless of the evidence provided, pointing to details that supported their viewpoint and disregarding anything contrary. Participants described studies supporting their pre-existing view as superior to those that contradicted it, in detailed and specific ways.

Writing about a study that seemed to undermine the deterrence effect, a death penalty proponent wrote, "The research didn't cover a long enough period of time", while an opponent's comment on the same study said, "No strong evidence to contradict the researchers has been presented". The results illustrated that people set higher standards of evidence for hypotheses that go against their current expectations. This effect, known as "disconfirmation bias", has been supported by other experiments.

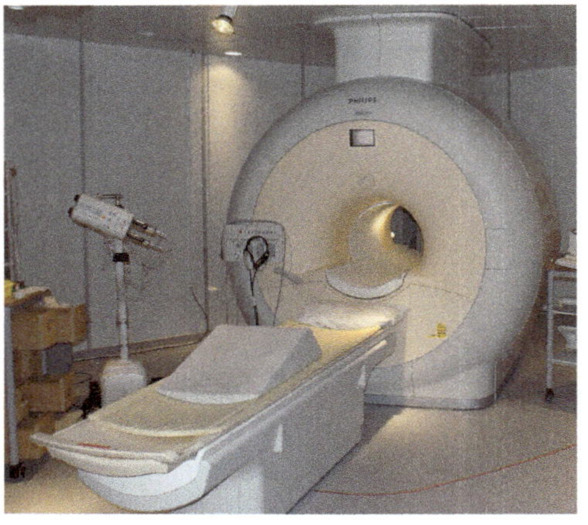

An MRI scanner allowed researchers to examine how the human brain deals with unwelcome information.

Another study of biased interpretation occurred during the 2004 U.S. presidential election and involved participants who reported having strong feelings about the candidates. They were shown apparently contradictory pairs of statements, either from Republican candidate George W. Bush, Democratic candidate John Kerry or a politically neutral public figure. They were also given further statements that made the apparent contradiction seem reasonable. From these three pieces of information, they had to decide whether or not each individual's statements were inconsistent. There were strong differences in these evaluations, with participants much more likely to interpret statements from the candidate they opposed as contradictory.

In this experiment, the participants made their judgments while in a magnetic resonance imaging (MRI) scanner which monitored their brain activity. As participants evaluated contradictory statements by their favored candidate, emotional centers of their brains were aroused. This did not happen with the statements by the other figures. The experimenters inferred that the different responses to the statements were not due to passive reasoning errors. Instead, the participants were actively reducing the cognitive dissonance induced by reading about their favored candidate's irrational or hypocritical behavior.

Biases in belief interpretation are persistent, regardless of intelligence level. Participants in an experiment took the SAT test (a college admissions test used in the United States) to assess their intelligence levels. They then read information regarding safety concerns for vehicles, and the experimenters manipulated the national origin of the car. American participants provided their opinion if the car should be banned on a six-point scale, where one indicated "definitely yes" and six indicated "definitely no." Participants firstly evaluated if they would allow a dangerous German car on American streets and a dangerous American car on German streets. Participants believed that the dangerous German car on American streets should be banned more quickly than the dangerous American car on German streets. There was no difference among intelligence levels at the rate participants would ban a car.

Biased interpretation is not restricted to emotionally significant topics. In another experiment, participants were told a story about a theft. They had to rate the evidential importance of statements arguing either for or against a particular character being responsible. When they hypothesized that character's guilt, they rated statements supporting that hypothesis as more important than conflicting statements.

Biased Memory

Even if people gather and interpret evidence in a neutral manner, they may still remember it selectively to reinforce their expectations. This effect is called "selective recall", "confirmatory memory" or "access-biased memory". Psychological theories differ in their predictions about selective recall. Schema theory predicts that information matching prior expectations will be more easily stored and recalled than information that does not match. Some alternative approaches say that surprising information stands out and so is memorable. Predictions from both these theories have been confirmed in different experimental contexts, with no theory winning outright.

In one study, participants read a profile of a woman which described a mix of introverted and extroverted behaviors. They later had to recall examples of her introversion and extroversion. One group was told this was to assess the woman for a job as a librarian, while a second group were told it was for a job in real estate sales. There was a significant difference between what these two groups recalled, with the "librarian" group recalling more examples of introversion and the "sales" groups recalling more extroverted behavior. A selective memory effect has also been shown in experiments that manipulate the desirability of personality types. In one of these, a group of participants were shown evidence that extroverted people are more successful than introverts. Another group were told the opposite. In a subsequent, apparently unrelated, study, they were asked to recall events from their lives in which they had been either introverted or extroverted. Each group of participants provided more memories connecting themselves with the more desirable personality type, and recalled those memories more quickly.

Changes in emotional states can also influence memory recall. Participants rated how they felt when they had first learned that O.J. Simpson had been acquitted of murder charges. They described their emotional reactions and confidence regarding the verdict one week, two months, and one year after the trial. Results indicated that participants' assessments for Simpson's guilt changed over time. The more that participants' opinion of the verdict had changed, the less stable were the participant's memories regarding their initial emotional reactions. When participants recalled their initial emotional reactions two months and a year later, past appraisals closely resembled current appraisals of emotion. People demonstrate sizable myside bias when discussing their opinions on controversial topics. Memory recall and construction of experiences undergo revision in relation to corresponding emotional states.

Myside bias has been shown to influence the accuracy of memory recall. In an experiment, widows and widowers rated the intensity of their experienced grief six months and five years after the deaths of their spouses. Participants noted a higher experience of grief at six months rather than at five years. Yet, when the participants were asked after five years how they had felt six months after the death of their significant other, the intensity of grief participants recalled was highly correlated with their current level of grief. Individuals appear to utilize their current emotional states to analyze how they must have felt when experiencing past events. Emotional memories are reconstructed by current emotional states.

One study showed how selective memory can maintain belief in extrasensory perception (ESP). Believers and disbelievers were each shown descriptions of ESP experiments. Half of each group were told that the experimental results supported the existence of ESP, while the others were told they did not. In a subsequent test, participants recalled the material accurately, apart from believers who had read the non-supportive evidence. This group remembered significantly less information and some of them incorrectly remembered the results as supporting ESP.

Related Effects

Polarization of Opinion

When people with opposing views interpret new information in a biased way, their views can move even further apart. This is called "attitude polarization". The effect was demonstrated by an experiment that involved drawing a series of red and black balls from one of two concealed "bingo baskets". Participants knew that one basket contained 60% black and 40% red balls; the other, 40% black and 60% red. The experimenters looked at what happened when balls of alternating color were drawn in turn, a sequence that does not favor either basket. After each ball was drawn, participants in one group were asked to state out loud their judgments of the probability that the balls were being drawn from one or the other basket. These participants tended to grow more confident with each successive draw—whether they initially thought the basket with 60% black balls or the one with 60% red balls was the more likely source, their estimate of the probability increased. Another group of participants were asked to state probability estimates only at the end of a sequence of drawn balls, rather than after each ball. They did not show the polarization effect, suggesting that it does not necessarily occur when people simply hold opposing positions, but rather when they openly commit to them.

A less abstract study was the Stanford biased interpretation experiment in which participants with strong opinions about the death penalty read about mixed experimental evidence. Twenty-three percent of the participants reported that their views had become more extreme, and this self-reported shift correlated strongly with their initial attitudes. In later experiments, participants also reported their opinions becoming more extreme in response to ambiguous information. However, comparisons of their attitudes before and after the new evidence showed no significant change, suggesting that the self-reported changes might not be real. Based on these experiments, Deanna Kuhn and Joseph Lao concluded that polarization is a real phenomenon but far from inevitable, only happening in a small minority of cases. They found that it was prompted not only by considering mixed evidence, but by merely thinking about the topic.

Charles Taber and Milton Lodge argued that the Stanford team's result had been hard to replicate because the arguments used in later experiments were too abstract or confusing to evoke an emotional response. The Taber and Lodge study used the emotionally charged topics of gun control and affirmative action. They measured the attitudes of their participants towards these issues before and after reading arguments on each side of the debate. Two groups of participants showed attitude polarization: those with strong prior opinions and those who were politically knowledgeable. In part of this study, participants chose which information sources to read, from a list prepared by the experimenters. For example, they could read the National Rifle Association's and the Brady Anti-Handgun Coalition's arguments on gun control. Even when instructed to be even-handed, participants were more likely to read arguments that supported their existing attitudes than arguments that did not. This biased search for information correlated well with the polarization effect.

The *backfire effect* is a name for the finding that, given evidence against their beliefs, people can reject the evidence and believe even more strongly. The phrase was first coined by Brendan Nyhan and Jason Reifler.

Persistence of Discredited Beliefs

[B]eliefs can survive potent logical or empirical challenges. They can survive and even be bolstered by evidence that most uncommitted observers would agree logically demands some weakening of such beliefs. They can even survive the total destruction of their original evidential bases.

—Lee Ross and Craig Anderson

Confirmation biases can be used to explain why some beliefs persist when the initial evidence for them is removed. This belief perseverance effect has been shown by a series of experiments using what is called the "debriefing paradigm": participants read fake evidence for a hypothesis, their attitude change is measured, then the fakery is exposed in detail. Their attitudes are then measured once more to see if their belief returns to its previous level.

A common finding is that at least some of the initial belief remains even after a full debriefing. In one experiment, participants had to distinguish between real and fake suicide notes. The feedback was random: some were told they had done well while others were told they had performed badly. Even after being fully debriefed, participants were still influenced by the feedback. They still thought they were better or worse than average at that kind of task, depending on what they had initially been told.

In another study, participants read job performance ratings of two firefighters, along with their responses to a risk aversion test. This fictional data was arranged to show either a negative or positive association: some participants were told that a risk-taking firefighter did better, while others were told they did less well than a risk-averse colleague. Even if these two case studies were true, they would have been scientifically poor evidence for a conclusion about firefighters in general. However, the participants found them subjectively persuasive. When the case studies were shown to be fictional, participants' belief in a link diminished, but around half of the original effect remained. Follow-up interviews established that the participants had understood the debriefing and taken it seriously. Participants seemed to trust the debriefing, but regarded the discredited information as irrelevant to their personal belief.

The *continued influence effect* is the tendency to believe previously learned misinformation even after it has been corrected. Misinformation can still influence inferences one generates after a correction has occurred.

Preference for Early Information

Experiments have shown that information is weighted more strongly when it appears early in a series, even when the order is unimportant. For example, people form a more positive impression of someone described as "intelligent, industrious, impulsive, critical, stubborn, envious" than when they are given the same words in reverse order. This *irrational primacy effect* is independent of the primacy effect in memory in which the earlier items in a series leave a stronger memory trace. Biased interpretation offers an explanation for this effect: seeing the initial evidence, people form a working hypothesis that affects how they interpret the rest of the information.

One demonstration of irrational primacy used colored chips supposedly drawn from two urns. Participants were told the color distributions of the urns, and had to estimate the probability of a chip being drawn from one of them. In fact, the colors appeared in a pre-arranged order. The first thirty draws favored one urn and the next thirty favored the other. The series as a whole was neutral, so rationally, the two urns were equally likely. However, after sixty draws, participants favored the urn suggested by the initial thirty.

Another experiment involved a slide show of a single object, seen as just a blur at first and in slightly better focus with each succeeding slide. After each slide, participants had to state their best guess of what the object was. Participants whose early guesses were wrong persisted with those guesses, even when the picture was sufficiently in focus that the object was readily recognizable to other people.

Illusory Association between Events

Illusory correlation is the tendency to see non-existent correlations in a set of data. This tendency was first demonstrated in a series of experiments in the late 1960s. In one experiment, participants read a set of psychiatric case studies, including responses to the Rorschach inkblot test. The participants reported that the homosexual men in the set were more likely to report seeing buttocks, anuses or sexually ambiguous figures in the inkblots. In fact the fictional case studies had been constructed so that the homosexual men were no more likely to report this imagery or, in one version of the experiment, were less likely to report it than heterosexual men. In a survey, a group of experienced psychoanalysts reported the same set of illusory associations with homosexuality.

Another study recorded the symptoms experienced by arthritic patients, along with weather conditions over a 15-month period. Nearly all the patients reported that their pains were correlated with weather conditions, although the real correlation was zero.

This effect is a kind of biased interpretation, in that objectively neutral or unfavorable evidence is interpreted to support existing beliefs. It is also related to biases in hypothesis-testing behavior. In judging whether two events, such as illness and bad weather, are correlated, people rely heavily on the number of *positive-positive* cases: in this example, instances of both pain and bad weather. They pay relatively little attention to the other kinds of observation (of no pain and/or good weather). This parallels the reliance on positive tests in hypothesis testing. It may also reflect selective recall, in that people may have a sense that two events are correlated because it is easier to recall times when they happened together.

Individual Differences

Myside bias was once believed to be associated with greater intelligence; however, studies have shown that myside bias can be more influenced by ability to rationally think as opposed to amount of intelligence. Myside bias can cause an inability to effectively and logically evaluate the opposite side of an argument. Studies have stated that myside bias is an absence of “active open-mindedness,” meaning the active search for why an initial idea may be wrong. Typically, myside bias is operationalized in empirical studies as the quantity of evidence used in support of their side in comparison to the opposite side.

A study has found individual differences in myside bias. This study investigates individual differences that are acquired through learning in a cultural context and are mutable. The researcher found important individual difference in argumentation. Studies have suggested that individual differences such as deductive reasoning ability, ability to overcome belief bias, epistemological understanding, and thinking disposition are significant predictors of the reasoning and generating arguments, counterarguments, and rebuttals.

A study by Christopher Wolfe and Anne Britt also investigated how participants' views of "what makes a good argument?" can be a source of myside bias that influence the way a person formulates his own arguments. The study investigated individual differences of argumentation schema and asked participants to write essays. The participants were randomly assigned to write essays either for or against their preferred side of an argument and were given research instructions that took either a balanced or an unrestricted approach. The balanced-research instructions directed participants to create a "balanced" argument, i.e., that included both pros and cons; the unrestricted-research instructions included nothing on how to create the argument.

Overall, the results revealed that the balanced-research instructions significantly increased the incidence of opposing information in arguments. These data also reveal that personal belief is not a *source* of myside bias; however, that those participants, who believe that a good argument is one that is based on facts, are more likely to exhibit myside bias than other participants. This evidence is consistent with the claims proposed in Baron's article—that people's opinions about what makes good thinking can influence how arguments are generated.

History

Francis Bacon

Informal observation

Before psychological research on confirmation bias, the phenomenon had been observed anecdotally throughout history. Beginning with the Greek historian Thucydides (c. 460 BC – c. 395 BC), who wrote of misguided treason in *The Peloponnesian War*; "... for it is a habit of mankind to entrust to careless hope what they long for, and to use sovereign reason to thrust aside what they do not fancy." Italian poet Dante Alighieri

(1265–1321), noted it in his famous work, the *Divine Comedy*, in which St. Thomas Aquinas cautions Dante upon meeting in Paradise, "opinion—hasty—often can incline to the wrong side, and then affection for one's own opinion binds, confines the mind." English philosopher and scientist Francis Bacon (1561–1626), in the *Novum Organum* noted that biased assessment of evidence drove "all superstitions, whether in astrology, dreams, omens, divine judgments or the like". He wrote:

The human understanding when it has once adopted an opinion ... draws all things else to support and agree with it. And though there be a greater number and weight of instances to be found on the other side, yet these it either neglects or despises, or else by some distinction sets aside or rejects.

In his essay (1897) "What Is Art?", Russian novelist Leo Tolstoy wrote,

I know that most men—not only those considered clever, but even those who are very clever, and capable of understanding most difficult scientific, mathematical, or philosophic problems—can very seldom discern even the simplest and most obvious truth if it be such as to oblige them to admit the falsity of conclusions they have formed, perhaps with much difficulty—conclusions of which they are proud, which they have taught to others, and on which they have built their lives.

Wason's research on hypothesis-testing

The term "confirmation bias" was coined by English psychologist Peter Wason. For an experiment published in 1960, he challenged participants to identify a rule applying to triples of numbers. At the outset, they were told that (2,4,6) fits the rule. Participants could generate their own triples and the experimenter told them whether or not each triple conformed to the rule.

While the actual rule was simply "any ascending sequence", the participants had a great deal of difficulty in finding it, often announcing rules that were far more specific, such as "the middle number is the average of the first and last". The participants seemed to test only positive examples—triples that obeyed their hypothesized rule. For example, if they thought the rule was, "Each number is two greater than its predecessor", they would offer a triple that fit this rule, such as (11,13,15) rather than a triple that violates it, such as (11,12,19).

Wason accepted falsificationism, according to which a scientific test of a hypothesis is a serious attempt to falsify it. He interpreted his results as showing a preference for confirmation over falsification, hence the term "confirmation bias". Wason also used confirmation bias to explain the results of his selection task experiment. In this task, participants are given partial information about a set of objects, and have to specify what further information they would need to tell whether or not a conditional rule ("If A, then B") applies. It has been found repeatedly that people perform badly on various forms of this test, in most cases ignoring information that could potentially refute the rule.

Klayman and Ha's Critique

A 1987 paper by Joshua Klayman and Young-Won Ha argued that the Wason experiments had not actually demonstrated a bias towards confirmation. Instead, Klayman and Ha interpreted the results in terms of a tendency to make tests that are consistent with the working hypothesis. They called this the "positive test strategy". This strategy is an example of a heuristic: a reasoning shortcut that is imperfect but easy to compute. Klayman and Ha used Bayesian probability and information theory as their standard of hypothesis-testing, rather than the falsificationism used by Wason. According to these ideas, each answer to a question yields a different amount of information, which depends on the person's prior beliefs. Thus a scientific test of a hypothesis is one that is expected to produce the most information. Since the information content depends on initial probabilities, a positive test can either be highly informative or uninformative. Klayman and Ha argued that when people think about realistic problems, they are looking for a specific answer with a small initial probability. In this case, positive tests are usually more informative than negative tests. However, in Wason's rule discovery task the answer—three numbers in ascending order—is very broad, so positive tests are unlikely to yield informative answers. Klayman and Ha supported their analysis by citing an experiment that used the labels "DAX" and "MED" in place of "fits the rule" and "doesn't fit the rule". This avoided implying that the aim was to find a low-probability rule. Participants had much more success with this version of the experiment.

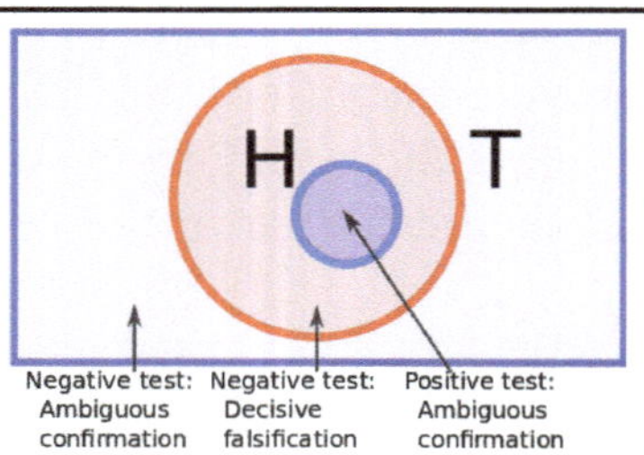

If the true rule (T) encompasses the current hypothesis (H), then positive tests (examining an H to see if it is T) will not show that the hypothesis is false.

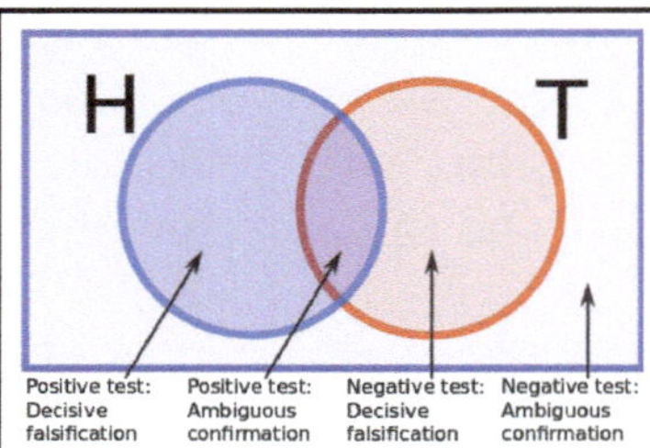

If the true rule (T) *overlaps* the current hypothesis (H), then either a negative test or a positive test can potentially falsify H.

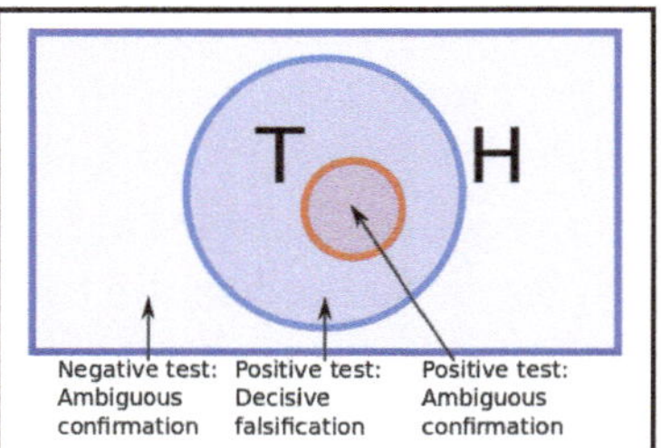

When the working hypothesis (H) includes the true rule (T) then positive tests are the *only* way to falsify H.

In light of this and other critiques, the focus of research moved away from confirmation versus falsification to examine whether people test hypotheses in an informative way, or an uninformative but positive way. The search for "true" confirmation bias led psychologists to look at a wider range of effects in how people process information.

Explanations

Confirmation bias is often described as a result of automatic, unintentional strategies rather than deliberate deception. According to Robert Maccoun, most biased evidence processing occurs through a combination of both "cold" (cognitive) and "hot" (motivated) mechanisms.

Cognitive explanations for confirmation bias are based on limitations in people's ability to handle complex tasks, and the shortcuts, called *heuristics*, that they use. For example, people may judge the reliability of evidence by using the *availability heuristic*—i.e., how readily a particular idea comes to mind. It is also possible that people can only focus on one thought at a time, so find it difficult to test alternative hypotheses in parallel. Another heuristic is the positive test strategy identified by Klayman and Ha, in which people test a hypothesis by examining cases where they expect a property or event to occur. This heuristic avoids the difficult or impossible task of working out how diagnostic each possible question will be. However, it is not universally reliable, so people can overlook challenges to their existing beliefs.

Motivational explanations involve an effect of desire on belief, sometimes called "wishful thinking". It is known that people prefer pleasant thoughts over unpleasant ones in a number of ways: this is called the "Pollyanna principle". Applied to arguments or sources of evidence, this could explain why desired conclusions are more likely to be believed true. According to experiments that manipulate the desirability of the conclusion, people demand a high standard of evidence for unpalatable ideas and a low standard for preferred ideas. In other words, they ask, "Can I believe this?" for some suggestions and, "Must I believe this?" for others. Although consistency is a desirable feature of attitudes, an excessive drive for consistency is another potential source of bias because it may prevent people from neutrally evaluating new, surprising information. Social psychologist Ziva Kunda combines the cognitive and motivational theories, arguing that motivation creates the bias, but cognitive factors determine the size of the effect.

Explanations in terms of cost-benefit analysis assume that people do not just test hypotheses in a disinterested way, but assess the costs of different errors. Using ideas from evolutionary psychology, James Friedrich suggests that people do not primarily aim at truth in testing hypotheses, but try to avoid the most costly errors. For example, employers might ask one-sided questions in job interviews because they are focused on weeding out unsuitable candidates. Yaacov Trope and Akiva Liberman's refinement of this theory assumes that people compare the two different kinds of error: accepting a false hypothesis or rejecting a true hypothesis. For instance, someone who underestimates a friend's honesty might treat him or her suspiciously and so undermine the friendship. Overestimating the friend's honesty may also be costly, but less so. In this case, it would be rational to seek, evaluate or remember evidence of their honesty in a biased way. When someone gives an initial impression of being introverted or extroverted, questions that match that impression come across as more empathic. This suggests that when talking to someone who seems to be an introvert, it is a sign of better social skills to ask, "Do you feel awkward in social situations?" rather than, "Do you like noisy parties?" The connection between confirmation bias and social skills was corroborated by a study of how college students get to know other people. Highly self-monitoring students, who are more sensitive to their environment and to social norms, asked more matching questions when interviewing a high-status staff member than when getting to know fellow students.

Psychologists Jennifer Lerner and Philip Tetlock distinguish two different kinds of thinking process. *Exploratory thought* neutrally considers multiple points of view and tries to anticipate all possible objections to a particular position, while *confirmatory thought* seeks to justify a specific point of view. Lerner and Tetlock say that when people expect to justify their position to others whose views they already know, they will tend to adopt a similar position to those people, and then use confirmatory thought to bolster their own credibility. However, if the external parties are overly aggressive or critical, people will disengage from thought altogether, and simply assert their personal opinions without justification. Lerner and Tetlock say that people only push themselves to think critically and logically when they know in advance they will need to explain themselves to others who are well-informed, genuinely interested in the truth, and whose views they don't already know. Because those conditions rarely exist, they argue, most people are using confirmatory thought most of the time.

Consequences

In finance

Confirmation bias can lead investors to be overconfident, ignoring evidence that their strategies will lose money. In studies of political stock markets, investors made more profit when they resisted bias. For example, participants who interpreted a candidate's debate performance in a neutral rather than partisan way were more likely to profit. To combat the effect of confirmation bias, investors can try to adopt a contrary viewpoint "for the sake of argument". In one technique, they imagine that their investments have collapsed and ask themselves why this might happen.

In physical and mental health

Raymond Nickerson, a psychologist, blames confirmation bias for the ineffective medical procedures that were used for centuries before the arrival of scientific medicine. If a patient recovered, medical authorities counted the treatment as successful, rather than looking for alternative explanations such as that the disease had run its natural course. Biased assimilation is a factor in the modern appeal of alternative medicine, whose proponents are swayed by positive anecdotal evidence but treat scientific evidence hyper-critically.

Cognitive therapy was developed by Aaron T. Beck in the early 1960s and has become a popular approach. According to Beck, biased information processing is a factor in depression. His approach teaches people to treat evidence impartially, rather than selectively reinforcing negative outlooks. Phobias and hypochondria have also been shown to involve confirmation bias for threatening information.

In Politics and Law

Mock trials allow researchers to examine confirmation biases in a realistic setting.

Nickerson argues that reasoning in judicial and political contexts is sometimes subconsciously biased, favoring conclusions that judges, juries or governments have already committed to. Since the evidence in a jury trial can be complex, and jurors often reach decisions about the verdict early on, it is reasonable to expect an attitude polarization effect. The prediction that jurors will become more extreme in their views as they see more evidence has been borne out in experiments with mock trials. Both inquisitorial and adversarial criminal justice systems are affected by confirmation bias.

Confirmation bias can be a factor in creating or extending conflicts, from emotionally charged debates to wars: by interpreting the evidence in their favor, each opposing party can become overconfident that it is in the stronger position. On the other hand, confirmation bias can result in people ignoring or misinterpreting the signs of an imminent or incipient conflict. For example, psychologists Stuart Sutherland and Thomas Kida have each argued that US Admiral Husband E. Kimmel showed confirmation bias when playing down the first signs of the Japanese attack on Pearl Harbor.

A two-decade study of political pundits by Philip E. Tetlock found that, on the whole, their predictions were not much better than chance. Tetlock divided experts into "foxes" who maintained multiple hypotheses, and "hedgehogs" who were more dogmatic. In general, the hedgehogs were much less accurate. Tetlock blamed their failure on confirmation bias—specifically, their inability to make use of new information that contradicted their existing theories.

In the 2013 murder trial of David Camm, the defense argued that Camm was charged for the murders of his wife and two children solely because of confirmation bias within the investigation. Camm was arrested three days after the murders on the basis of faulty evidence. Despite the discovery that almost every piece of evidence on the probable cause affidavit was inaccurate or unreliable, the charges were not dropped against him. A sweatshirt found at the crime was subsequently discovered to contain the DNA of a convicted felon, his prison nickname, and his department of corrections number. Investigators looked for Camm's DNA on the sweatshirt, but failed to investigate any other pieces of evidence found on it and the foreign DNA was not run through CODIS until 5 years after the crime. When the second suspect was discovered, prosecutors charged them as co-conspirators in the crime despite finding no evidence linking the two men. Camm was acquitted of the murders.

In the Paranormal

One factor in the appeal of alleged psychic readings is that listeners apply a confirmation bias which fits the psychic's statements to their own lives. By making a large number of ambiguous statements in each sitting, the psychic gives the client more opportunities to find a match. This is one of the techniques of cold reading, with which a psychic can deliver a subjectively impressive reading without any prior information about the client. Investigator James Randi compared the transcript of a reading to the client's report of what the psychic had said, and found that the client showed a strong selective recall of the "hits".

As a striking illustration of confirmation bias in the real world, Nickerson mentions numerological pyramidology: the practice of finding meaning in the proportions of the Egyptian pyramids. There are many different length measurements that can be made of, for example, the Great Pyramid of Giza and many ways to combine or manipulate them. Hence it is almost inevitable that people who look at these numbers selectively will find superficially impressive correspondences, for example with the dimensions of the Earth.

In Science

A distinguishing feature of scientific thinking is the search for falsifying as well as confirming evidence. However, many times in the history of science, scientists have resisted new discoveries by selectively interpreting or ignoring unfavorable data. Previous research has shown that the assessment of the quality of scientific studies seems to be particularly vulnerable to confirmation bias. It has been found several times that scientists rate studies that report findings consistent with their prior beliefs more favorably than studies reporting findings inconsistent with their previous beliefs. However, assuming that the research question is relevant, the experimental design adequate and the data are clearly and comprehensively described, the found results should be of importance to the scientific community and should not be viewed prejudicially, regardless of whether they conform to current theoretical predictions.

In the context of scientific research, confirmation biases can sustain theories or research programs in the face of inadequate or even contradictory evidence; the field of parapsychology has been particularly affected.

An experimenter's confirmation bias can potentially affect which data are reported. Data that conflict with the experimenter's expectations may be more readily discarded as unreliable, producing the so-called file drawer effect. To combat this tendency, scientific training teaches ways to prevent bias. For example, experimental design of randomized controlled trials (coupled with their systematic review) aims to minimize sources of bias. The social process of peer review is thought to mitigate the effect of individual scientists' biases, even though the peer review process itself may be susceptible to such biases. Confirmation bias may thus be especially harmful to objective evaluations regarding nonconforming results since biased individuals may regard opposing evidence to be weak in principle and give little serious thought to revising their beliefs. Scientific innovators often meet with resistance from the scientific community, and research presenting controversial results frequently receives harsh peer review.

In Self-image

Social psychologists have identified two tendencies in the way people seek or interpret information about themselves. *Self-verification* is the drive to reinforce the existing self-image and *self-enhancement* is the drive to seek positive feedback. Both are served by confirmation biases. In experiments where people are given feedback that conflicts with their self-image, they are less likely to attend to it or remember it than when given self-verifying feedback. They reduce the impact of such information by interpreting it as unreliable. Similar experiments have found a preference for positive feedback, and the people who give it, over negative feedback.

Ambidextrous Organization

Organizational ambidexterity refers to an organization's ability to be efficient in its management of today's business and also adaptable for coping with tomorrow's changing demand. Just as being ambidextrous means being able to use both the left and right hand equally, organizational ambidexterity requires the organizations to use both exploration and exploitation techniques to be successful.

Origin and Development

Organizational ambidexterity was defined as an organization's ability to be aligned and efficient in its management of today's business demands as well as being adaptive to changes in the environment at the same time. This term of organizational ambidexterity was first used by Duncan, however, it was March that had been credited for developing and generating greater interest in this concept, especially in the late 20th and early 21st century. Ambidexterity in an organization is achieved by balancing exploration and exploitation, which allows the organization to be creative and adaptable, while also continuing to rely on more traditional, proven methods of business. Exploration includes things such as search, variation, risk taking, experimentation, flexibility, discovery or innovation, whereas exploitation includes such things as refinement, choice, production, efficiency, selection, implementation, and execution. Companies that focus only on exploration face the risk of wasting resources on ideas that may not prove useful or never be developed. On the other hand, companies that focus only on exploitation may accept status quo performance and products and fail to reach optimal levels of success.

Organizational ambidexterity is defined broadly, and several other terms are also highly related or similar to the construct of ambidextrous organization, including organizational learning, technological innovation, organizational adaptation, strategic management, and organizational design. Things such as reconciling exploitation and exploration, the simultaneity of induced and autonomous strategy processes, synchronizing incremental and discontinuous innovation, and balancing search and stability also tend to refer to the same underlying construct as ambidextrous organization.

Whereas earlier studies regarded the trade-offs between exploration and exploitation to be insurmountable, more recent research has paid attention to a range of organizational solutions to engender the existence of ambidexterity. One recent hot research topic in this area focused on the leadership characteristics that enable organizations to manage the contradictions that they face and achieve ambidexterity, which is the origin of the concept 'ambidextrous leadership'. Several antecedents, outcomes of organizational ambidexterity as well as related moderators have also been identified in the existing literature. These topics are introduced in later sections.

Two Models

Organizational ambidexterity can be considered primarily from two angles. One is architectural or structural ambidexterity, which uses dual organizational structures and strategies to differentiate efforts towards exploitation and exploration. Structural ambidexterity includes dual parts, with one part focusing on exploitation and the other focusing on exploration. It's also known as the spatial separation of the dual strategies concepts outlined above. The other approach is contextual ambidexterity, which uses behavioral and social means to integrate exploitation and exploration at the organizational unit level. Contextual ambidexterity is a balanced type that takes a mid-level position between exploitation and exploration, also known as parallel structures or hybrid strategies.

Although both angles are related to the theme of organizational ambidexterity, they strongly differ in the way how they are configured. There has always been a debate of which of the two different approaches is right. The dual type allows both discrete orientations and fit across dimensions at the unit level but creates a misfit between organizational units. Some researchers argued that inconsistent structures may lead to low performance. There are also some researchers trying to investigate the external and internal contingencies under which different configurations can be found. One factor would be the speed and type of technological change that organizations confront. On the other hand, the balanced type (i.e. contextual ambidexterity) is consistent with the systems approach of fit across multiple dimensions, but contradicts the opinion that organizational choice is discrete. In an environment where changes are slow, there will be sufficient time to react to the radical changes by constructing dual structures and strategies. However, in a high-competitive environment, balanced structures may be better prepared to deal with the constant need for alignment. In future studies, the different organizational ambidexterity configurations can be compared to find a better solution for dealing with the exploitation and exploration paradox.

Antecedents

Ambidexterity is often considered a capacity or skill and can therefore be facilitated by either the organization as a whole or managers / CEOs. From the structural ambidexterity's view, organizations can solve the paradox by temporarily cycling through periods of exploitation and periods of exploration. From the other point of view (contextual ambidexterity), firms ought to address exploitation and exploration simultaneously and internally to achieve the goal ambidexterity. Contextual ambidexterity is more difficult to achieve than structural ambidexterity because managing two inconsistent alignments within an organization simultaneously is far more complex than managing one consistent strategy after another. Thus most studies on how to build ambidextrous organizations in literature focused on contextual ambidexterity rather than structural ambidexterity.

The role of leaders (or managers) is always highlighted towards building an ambidextrous organization. Several recommendations have been made to organizations on how to achieve contextual ambidexterity, including using of meta-routines and job-enrichment schemes, building trust with supervisees, being supportive, using complex behavioral repertoires, and the creation of a shared vision.

In addition, several characteristics of organizational culture are also emphasized on building an ambidextrous organization. Successful organizations should be able to balance the hard elements (discipline and stretch) and the soft elements (support and trust) in their organizational contexts. It was also suggested that establishing shared goals, developing a collective identity, creating a culture of support, and giving personal meaning to individuals' contributions to the overall purpose of an organization all contributes to ambidexterity. A decentralized structure and a culture of excellence, recruitment and training are also important aspects of establishing an ambidextrous organization.

Outcomes

Ambidexterity is beneficial to organizations in many aspects. As it is the ability to keep a balance between explorative and exploitative processes, the most core outcome of ambidexterity is innovation because innovation needs both explorative and exploitative aspects. Innovation is defined as "the sequence of activities by which a new element is introduced into a social unit, with the intention of benefiting the unit, some part of it, or the wider society" (West & Farr, 1990). The new element need not be entirely novel or unfamiliar to members of the unit, but it must involve some discernible change or challenge to the status quo (West & Farr, 1990). Most theoretical models of innovation differentiate at least two processes: idea generation and idea implementation (e.g., Amabile, 1988; West, 2002). The generation phase is closely linked to explorative activities while the implementation phase is highly linked to exploitative activities. An ambidextrous organization is able to pursue innovation (creating new products/services) while also maintaining itself through the continued use of proven techniques/products.

In addition, ambidexterity can promote some other positive organizational outcomes besides innovation. It has been proved in literature that the interaction between explorative and exploitative innovation strategies (in other words, ambidexterity) is positively related to sales growth rate, and imbalance between explorative and exploitative innovation strategies is negatively related to sales growth rate (He & Wong, 2004). Various organizations have been able to overcome organizational challenges and become more innovative because of ambidexterity. A study looking at 41 businesses found that ambidexterity was highly correlated with performance. Similarly, another study of 34 high-tech organizations showed that their ability to simultaneously execute exploration and exploitation was associated with higher performance (Chandrasekaran et al. 2012).

Companies such as Apple, General Radio, Nordstrom and British Airways have all had continued success throughout the years because of their capacity for ambidexterity. From 1981 to 1986, British Airways experienced increased profits and customer satisfaction. The top executives of British Airways credited the formation of a more ambidextrous culture and leadership with the company's improved performance.

Related Moderators

Moderators exist in the relationship between organizational ambidexterity and organizational outcomes. Environmental dynamism and competitiveness moderate the relationship between exploitation/ exploration and performance (Levinthal & March, 1993; Lewin, Long & Carroll, 1999). Empirical studies also showed that pursuing exploratory innovation is more effective in dynamic environments, whereas pursuing exploitative innovation is more beneficial to a unit's financial performance in more competitive environments (Jansen, van den Bosch, & Volberda, 2006). Although they were not directly testing an ambidextrous orientation, results indicated a positive performance effect of simultaneously pursuing exploitative and exploratory innovation under high dynamic and competitive environments. The effects of exploitative, explorative and balanced corporate alignment activities on performance were compared under varying environmental conditions (Raisch & Hotz, 2008). The construct of "environmental munificence" was developed to reflect an organization's opportunities and dynamism

(Zahra, 1993). They found that, although exploration was positively related to performance under high environmental munificence, a balanced orientation failed to significantly affect performance in times of low environmental munificence (Raisch & Hotz, 2008).

Market orientation was also proved to moderate the relationship between exploitation/exploration and performance. Market orientation was defined as "the capability to generate, disseminate, and respond to intelligence pertaining to current and future customers" (Kohli & Jaworski, 1990). A longitudinal study by Kyriakopoulos and Moorman (2004) showed that market orientation positively moderates the impact of pursuing high levels of exploitative and exploratory marketing strategies on new product performance; however, firms that pursue an ambidextrous orientation without strong market orientation display a significant reduction in new product financial performance.

Another factor that may moderate ambidexterity's effect on performance is firm resources (Kyriakopoulos & Moorman, 2004). Firms with rich resources have the ability to exploit and explore simultaneously, whereas firms with less resources may not be able to afford such a complex strategy. Similarly, Lubatkin et al. stated that small firms "lack the amount of slack resources and the kind of hierarchical administration systems that can help or impede larger firms in managing their contradictory knowledge processes and, thus, affect the attainment of ambidexterity" (p. 647). This idea was supported by empirical evidence that small firms may benefit more from a one-sided orientation than from mixed strategies (Ebben & Johnson, 2005).

Boundary conditions were also addressed on choosing between structural and contextual ambidexterity. For example, spatial separation was suggested as an appropriate solution for environments characterized by long periods of stability, disrupted by rare events of discontinuous change. Research also found that firms operating in dynamic competitive environments rely on contextual ambidexterity rather than developing spatially separated units (Jansen, van den Bosch, & Volberda, 2005).

Levels of Ambidexterity

The functional definition of ambidexterity was originally used to describe organizations, but recently this concept was extended to multiple organizational levels, including individuals, teams, and leaders. On the most general level, the concept of ambidexterity implies successfully managing the dichotomy of explorative variability creation and exploitative variability reduction.

Whenever there are needs to be both explorative and exploitive, conflict occurs (Bledow, Frese, Anderson, Erez, & Farr, 2009). That's when ambidexterity is necessary. Actually, regulating the conflicting demands of innovation is not only a challenge for the upper echelon of an organization but also a phenomenon that spans all levels of an organization. Employees as individuals, collectives of employees such as work teams, and the organization as a whole all have to find strategies to deal with conflicting demands in order to succeed in innovation and adaption to changing markets. Some examples of strategies and tactics that could be implemented at all three levels of analysis were also listed out (Bledow et al., 2009). These examples are presented in Table 1, including a separation strategy (in the Separation column) or an integration strategy (in the last two columns).

Ambidextrous Leadership

Recently the focus on organizational ambidexterity has become more concentrated on how leaders act ambidextrously to achieve organizational ambidexterity. Senior managers may be the key for facilitating the context and social base for ambidexterity. Noting that ambidextrous organizations require significant amounts of mobilization, coordination, and integration activities to maintain both exploitation and exploration, informal and social integration of the senior team as well as the cross-functional interfaces of the formal organization contribute to the success of organizational ambidexterity significantly (Jansen, Tempelaar, van den Bosch, & Volberda, 2009). A recent model of ambidexterity and leadership suggests that CEOs and top management teams (TMT) play an integral role in establishing ambidexterity in small-to-medium-sized organizations (Cao, Simsek, & Zhang, 2010). The model suggests TMTs are where processes such as information sharing and knowledge processing, which influence ambidexterity, are likely to occur. Furthermore, it is the CEO who has the ability to access the most valuable and diverse information needed to avoid separation of explorative and exploitative behaviors. The greater the interface between TMTs and CEOs in small-to-medium-sized organizations, the greater the amount of ambidexterity.

The concept of ambidexterity was first formally introduced into the leadership area by the Rosing, Frese and Bausch (2011) paper, holding the idea that leaders should be able to lead their team to match the complexity and the pace of innovation (Ancona, Goodman, Lawrence, & Tushman, 2001). Ambidextrous leadership was defined as the leaders' ability to foster both explorative and exploitative behaviors in followers by increasing or reducing variance in their behavior and flexibly switching between those behaviors (Rosing et al., 2011).

The construct of ambidextrous leadership has also been linked to the combination of leadership styles (Jansen et al., 2009). Leaders who are transformational encourage "out of the box thinking", information sharing and question assumptions.Transformational leaders promote exploration and innovative thinking. Transactional leaders focus on making incremental improvements and making the best use of existing process. The transactional leadership style promotes exploitative behaviors. An ambidextrous leader is able to switch back and forth between transformation/exploration and transaction/exploitation as needed, in other words, being able to switch between different leadership styles at the appropriate time, in order to foster innovation and then implement plans. Ambidextrous leadership consists of three elements (1) opening leader behaviors to foster exploration, (2) closing leader behaviors to foster exploitation, (3) and the temporal flexibility to switch between both as the situation requires (Rosing et al., 2011). Opening leadership behaviors include: allowing for multiple ways to accomplish a task, experimentation and errors, whereas closing behaviors include; monitoring routines, sticking to plans and minimizing errors.

Controversy and Future Directions

Some scholars as well as practitioners have argued that established companies simply lack the flexibility to explore new territories. One contributing reason could be the so-called success trap (i.e. the focus on their, historically successful, current business activities) (Walrave, Van Oorschot, & Romme, 2011). A possible solution for big companies is to adopt a venture capital model - funding exploratory expeditions but otherwise not interfering too much with their operations. Another suggestion is for the use of cross-functional teams to achieve breakthrough innovations. Still others have suggested that a company may be able to alternate between different organizational models, focusing on exploitation and exploration at different time periods. For example, in a study of biotechnology firms it is shown how an organization's management control system can be adjusted periodically to achieve this changing focus on exploitation and exploration. Researchers also debate if ambidexterity can be attained because exploration and exploitation tend to emerge from contradictory information and knowledge inputs and because success due to exploration/exploitation tends to be self-reinforcing leading to the use of the same methods in the future. An empirical study of ambidexterity in organizations (He & Wong, 2004) further cautions that very low levels of both exploration and exploitation are not sufficient to contribute to superior firm performance.

Ambidexterity can also be hampered by employee desire to receive organizational rewards. If organizations base their evaluation and reward systems on the completion of routine tasks, employees will be more likely to conform to standardized procedures. To avoid hindering ambidexterity, organizations should be sure to reward innovative thinking as well as completion of routine tasks.

Despite the controversy surrounding the possibility of organizational ambidexterity, it is likely to continue to be a highly researched concept in the future. Future research is likely to focus on the role of ambidextrous leaders, ambidextrous teams and the social context of organizational ambidexterity. (Rosing et al., 2011).

Disruptive Innovation

A disruptive innovation is an innovation that creates a new market and value network and eventually disrupts an existing market and value network, displacing established market leading firms, products and alliances. The term was defined and phenomenon analyzed by Clayton M. Christensen beginning in 1995. In the early 2000s, "significant societal impact" has also been used as an aspect of disruptive innovation.

The free, online encyclopedia Wikipedia was a disruptive innovation that had a major impact on both the traditional, for-profit printed paper encyclopedia market (e.g., *Encyclopedia Britannica*) and the for-profit digital encyclopedia market (e.g., Encarta). The English Wikipedia provides over 5 million articles for free; in contrast, a $1,000 set of Britannica volumes had 120,000 articles.

Not all innovations are disruptive, even if they are revolutionary. For example, the first automobiles in the late 19th century were not a disruptive innovation, because early automobiles were expensive luxury items that did not disrupt the market for horse-drawn vehicles. The market for transportation essentially remained intact until the debut of the lower-priced Ford Model T in 1908. The *mass-produced* automobile was a disruptive innovation, because it changed the transportation market, whereas the first thirty years of automobiles did not.

Disruptive innovations tend to be produced by outsiders and entrepreneurs, rather than existing market-leading companies. The business environment of market leaders does not allow them to pursue disruptive innovations when they first arise, because they are not profitable enough at first and because their development can take scarce resources away from sustaining innovations (which are needed to compete against current competition). A disruptive process can take longer to develop than by the conventional approach and the risk associated to it is higher than the other more incremental or evolutionary forms of innovations, but once it is deployed in the market, it achieves a much faster penetration and higher degree of impact on the established markets.

History and Usage of the Term

The term disruptive technologies was coined by Clayton M. Christensen and introduced in his 1995 article *Disruptive Technologies: Catching the Wave*, which he cowrote with Joseph Bower. The article is aimed at management executives who make the funding or purchasing decisions in companies, rather than the research community. He describes the term further in his book *The Innovator's Dilemma*. *Innovator's Dilemma* explored the cases of the disk drive industry (which, with its rapid generational change, is to the study of business what fruit flies are to the study of genetics, as Christensen was advised in the 1990s) and the excavating equipment industry (where hydraulic actuation slowly displaced cable-actuated movement). In his sequel with Michael E. Raynor, *The Innovator's Solution*, Christensen replaced the term *disruptive technology* with *disruptive innovation* because he recognized that few technologies are intrinsically disruptive or sustaining in character; rather, it is the *business model* that the technology enables that creates the disruptive impact. However, Christensen's evolution from a technological focus to a business-modelling focus is central to understanding the evolution of business at the market or industry level. Christensen and Mark W. Johnson, who cofounded the management consulting firm Innosight, described the dynamics of "business model innovation" in the 2008 *Harvard Business Review* article "Reinventing Your Business Model". The concept of disruptive technology continues a long tradition of identifying radical technical change in the study of innovation by economists, and the development of tools for its management at a firm or policy level.

In the late 1990s, the automotive sector began to embrace a perspective of "constructive disruptive technology" by working with the consultant David E. O'Ryan, whereby the use of current off-the-shelf technology was integrated with newer innovation to create what he called "an unfair advantage". The process or technology change as a whole had to be "constructive" in improving the current method of manufacturing, yet disruptively impact the whole of the business case model, resulting in a significant reduction of waste, energy, materials, labor, or legacy costs to the user.

In keeping with the insight that what matters economically is the business model, not the technological sophistication itself, Christensen's theory explains why many disruptive innovations are *not* "advanced technologies", which the technology mudslide hypothesis would lead one to expect. Rather, they are often novel combinations of existing off-the-shelf components, applied cleverly to a small, fledgling value network.

Theory

Christensen defines a disruptive innovation as a product or service designed for a new set of customers.

"Generally, disruptive innovations were technologically straightforward, consisting of off-the-shelf components put together in a product architecture that was often simpler than prior approaches. They offered less of what customers in established markets wanted and so could rarely be initially employed there. They offered a different package of attributes valued only in emerging markets remote from, and unimportant to, the mainstream."

Christensen argues that disruptive innovations can hurt successful, well-managed companies that are responsive to their customers and have excellent research and development. These companies tend to ignore the markets most susceptible to disruptive innovations, because the markets have very tight profit margins and are too small to provide a good growth rate to an established (sizable) firm. Thus, disruptive technology provides an example of an instance when the common business-world advice to "focus on the customer" (or "stay close to the customer", or "listen to the customer") can be strategically counterproductive.

While Christensen argued that disruptive innovations can hurt successful, well-managed companies, O'Ryan countered that "constructive" integration of existing, new, and forward-thinking innovation could improve the economic benefits of these same well-managed companies, once decision-making management understood the systemic benefits as a whole.

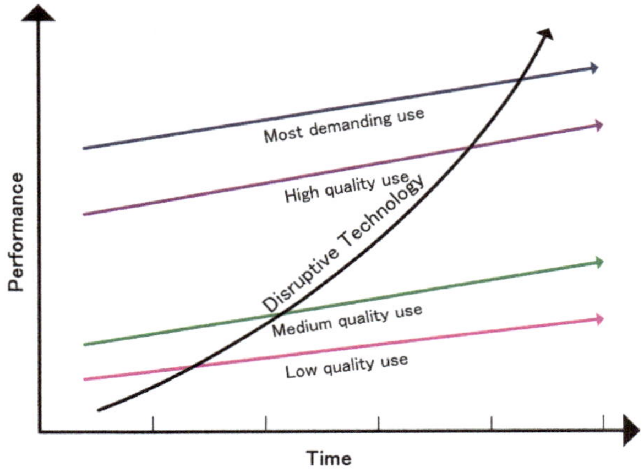

How low-end disruption occurs over time.

Christensen distinguishes between "low-end disruption", which targets customers who do not need the full performance valued by customers at the high end of the market, and "new-market disruption", which targets customers who have needs that were previously unserved by existing incumbents.

"Low-end disruption" occurs when the rate at which products improve exceeds the rate at which customers can adopt the new performance. Therefore, at some point the performance of the product overshoots the needs of certain customer segments. At this point, a disruptive technology may enter the market and provide a product that has lower performance than the incumbent but that exceeds the requirements of certain segments, thereby gaining a foothold in the market.

In low-end disruption, the disruptor is focused initially on serving the least profitable customer, who is happy with a good enough product. This type of customer is not willing to pay premium for enhancements in product functionality. Once the disruptor has gained a foothold in this customer segment, it seeks to improve its profit margin. To get higher profit margins, the disruptor needs to enter the segment where the customer is willing to pay a little more for higher quality. To ensure this quality in its product, the disruptor needs to innovate. The incumbent will not do much to retain its share in a not-so-profitable segment, and will move up-market and focus on its more attractive customers. After a number of such encounters, the incumbent is squeezed into smaller markets than it was previously serving. And then, finally, the disruptive technology meets the demands of the most profitable segment and drives the established company out of the market.

"New market disruption" occurs when a product fits a new or emerging market segment that is not being served by existing incumbents in the industry.

The extrapolation of the theory to all aspects of life has been challenged, as has the methodology of relying on selected case studies as the principal form of evidence. Jill Lepore points out that some companies identified by the theory as victims of disruption a decade or more ago, rather than being defunct, remain dominant in their industries today (including Seagate Technology, U.S. Steel, and Bucyrus). Lepore questions whether the theory has been oversold and misapplied, as if it were able to explain everything in every sphere of life, including not just business but education and public institutions.

Disruptive Technology

In 2009, Milan Zeleny described high technology as disruptive technology and raised the question of what is being disrupted. The answer, according to Zeleny, is the *support network* of high technology. For example, introducing electric cars disrupts the support network for gasoline cars (network of gas and service stations). Such disruption is fully expected and therefore effectively resisted by support net owners. In the long run, high (disruptive) technology bypasses, upgrades, or replaces the outdated support network.

Technology, being a form of social relationship, always evolves. No technology remains fixed. Technology starts, develops, persists, mutates, stagnates, and declines, just like living organisms. The evolutionary life cycle occurs in the use and development of any technology. A new high-technology core emerges and challenges existing technology support nets (TSNs), which are thus forced to coevolve with it. New versions of the core are designed and fitted into an increasingly appropriate TSN, with smaller and smaller high-technology effects. High technology becomes regular technology, with more efficient versions fitting the same support net. Finally, even the efficiency gains diminish, emphasis shifts to product tertiary attributes (appearance, style), and technology becomes TSN-preserving appropriate technology. This technological equilibrium state becomes established and fixated, resisting being interrupted by a technological mutation; then new high technology appears and the cycle is repeated.

Regarding this evolving process of technology, Christensen said:

"The technological changes that damage established companies are usually not radically new or difficult from a technological point of view. They do, however, have two important characteristics: First, they typically present a different package of performance attributes—ones that, at least at the outset, are not valued by existing customers. Second, the performance attributes that existing customers do value improve at such a rapid rate that the new technology can later invade those established markets."

Joseph Bower explained the process of how disruptive technology, through its requisite support net, dramatically transforms a certain industry.

"When the technology that has the potential for revolutionizing an industry emerges, established companies typically see it as unattractive: it's not something their mainstream customers want, and its projected profit margins aren't sufficient to cover big-company cost structure. As a result, the new technology tends to get ignored in favor of what's currently popular with the best customers. But then another company steps in to bring the innovation to a new market. Once the disruptive technology becomes established there, smaller-scale innovation rapidly raise the technology's performance on attributes that mainstream customers' value."

The automobile was high technology with respect to the horse carriage; however, it evolved into technology and finally into appropriate technology with a stable, unchanging TSN. The main high-technology advance in the offing is some form of electric car—whether the energy source is the sun, hydrogen, water, air pressure, or traditional charging outlet. Electric cars preceded the gasoline automobile by many decades and are now returning to replace the traditional gasoline automobile.

Milan Zeleny described the above phenomenon. He also wrote that:

"Implementing high technology is often resisted. This resistance is well understood on the part of active participants in the requisite TSN. The electric car will be resisted by gas-station operators in the same way automated teller machines (ATMs) were resisted by bank tellers and automobiles by horsewhip makers. Technology does not qualitatively restructure the TSN and therefore will not be resisted and never has been resisted. Middle management resists business process reengineering because BPR represents a direct assault on the support net (coordinative hierarchy) they thrive on. Teamwork and multi-functionality is resisted by those whose TSN provides the comfort of narrow specialization and command-driven work."

High-technology Effects

High technology is a technology core that changes the very architecture (structure and organization) of the components of the technology support net. High technology therefore transforms the qualitative nature of the TSN's tasks and their relations, as well as their requisite physical, energy, and information flows. It also affects the skills required, the roles played, and the styles of management and coordination—the organizational culture itself.

This kind of technology core is different from regular technology core, which preserves the qualitative nature of flows and the structure of the support and only allows users to perform the same tasks in the same way, but faster, more reliably, in larger quantities, or more efficiently. It is also different from appropriate technology core, which preserves the TSN itself with the purpose of technology implementation and allows users to do the same thing in the same way at comparable levels of efficiency, instead of improving the efficiency of performance.

As for the difference between high technology and low technology, Milan Zeleny once said:

"The effects of high technology always breaks the direct comparability by changing the system itself, therefore requiring new measures and new assessments of its productivity. High technology cannot be compared and evaluated with the existing technology purely on the basis of cost, net present value or return on investment. Only within an unchanging and relatively stable TSN would such direct financial comparability be meaningful. For example, you can directly compare a manual typewriter with an electric typewriter, but not a typewriter with a word processor. Therein lies the management challenge of high technology."

However, not all modern technologies are high technologies. They have to be used as such, function as such, and be embedded in their requisite TSNs. They have to empower the individual because only through the individual can they empower knowledge. Not all information technologies have integrative effects. Some information systems are still designed to improve the traditional hierarchy of command and thus preserve and entrench the existing TSN. The administrative model of management, for instance, further aggravates the division of task and labor, further specializes knowledge, separates management from workers, and concentrates information and knowledge in centers.

As knowledge surpasses capital, labor, and raw materials as the dominant economic resource, technologies are also starting to reflect this shift. Technologies are rapidly shifting from centralized hierarchies to distributed networks. Nowadays knowledge does not reside in a super-mind, super-book, or super-database, but in a complex relational pattern of networks brought forth to coordinate human action.

Practical Example of Disruption

In the practical world, the popularization of personal computers illustrates how knowledge contributes to the ongoing technology innovation. The original centralized concept (one computer, many persons) is a knowledge-defying idea of the prehistory of computing, and its inadequacies and failures have become clearly apparent. The era of personal computing brought powerful computers "on every desk" (one person, one computer). This short transitional period was necessary for getting used to the new computing environment, but was inadequate from the vantage point of producing knowledge. Adequate knowledge creation and management come mainly from networking and distributed computing (one person, many computers). Each person's computer must form an access point to the entire computing landscape or ecology through the Internet of other computers, databases, and mainframes, as well as production, distribution, and retailing facilities, and the like. For the first time, technology empowers individuals rather than external hierarchies. It transfers influence and power where it optimally belongs: at the loci of the useful knowledge. Even though hierarchies and bureaucracies do not innovate, free and empowered individuals do; knowledge, innovation, spontaneity, and self-reliance are becoming increasingly valued and promoted.

Examples

Category	Disruptive innovation	Market disrupted by innovation	Notes
Academia	Wikipedia	Traditional encyclopedias	Traditional, for-profit general encyclopedias with articles written by paid experts have been displaced by Wikipedia, an online encyclopedia which is written and edited by volunteer editors. Former market leader *Encyclopædia Britannica* ended its print production after 244 years in 2012. *Britannica*'s price of over $1000, its physical size of dozens of hard-bound volumes, its weight of over 100 pounds, its number of articles (about 120,000) and its update cycles lasting a year or longer made it unable to compete with Wikipedia, which provides free, online access to over 5 million articles which are updated every day. Wikipedia not only disrupted printed paper encyclopedias; it also disrupted digital encyclopedias. Microsoft's Encarta, a 1993 entry into professionally edited digital encyclopedias, was once a major rival to *Britannica* but was discontinued in 2009. Wikipedia's free access, online accessibility on computers and smartphones, unlimited size and instant updates are some of the challenges faced by for-profit competition in the encyclopedia market.
Communication	Telephony	Telegraphy	When Western Union declined to purchase Alexander Graham Bell's telephone patents for $100,000, their highest-profit market was long-distance telegraphy. Telephones were only useful at that time for very local calls. Short-distance telegraphy barely existed as a market segment, which explains Western Union's decision to not enter the emerging telephone market. However, telephones quickly displaced telegraphs, as telephones offered much greater communication capacity than telegraphs.

<table>
<tr><td rowspan="5">Computing hardware</td><td>Minicomputers</td><td>Mainframes</td><td rowspan="2">Minicomputers were originally presented as an inexpensive alternative to mainframes and mainframe manufacturers did not consider them a serious threat in their market. Eventually, the market for minicomputers (led by Seymor Cray—daisy chaining his minisupercomputers) became much larger than the market for mainframes.</td></tr>
<tr><td>Personal computers</td><td>Minicomputers, Workstations. Word processors, Lisp machines</td></tr>
<tr><td>Pocket calculator</td><td>3.5 standard calculator</td><td>Equivalent computing performance and portable</td></tr>
<tr><td>Digital calculator</td><td>Mechanical calculator</td><td>Facit AB used to dominate the European market for calculators, but did not adapt digital technology, and failed to compete with digital competitors.</td></tr>
<tr><td>Smartphones</td><td>Personal computers, laptops, PDAs</td><td>Smartphones and tablets are more portable than traditional PCs and laptops.</td></tr>
<tr><td rowspan="4">Data storage</td><td>8 inch floppy disk drive</td><td>14 inch hard disk drive</td><td rowspan="4">The floppy disk drive market has had unusually large changes in market share over the past fifty years. According to Clayton M. Christensen's research, the cause of this instability was a repeating pattern of disruptive innovations. For example, in 1981, the old 8 inch drives (used in mini computers) were "vastly superior" to the new 5.25 inch drives (used in desktop computers).

However, 8 inch drives were not affordable for the new desktop machines. The simple 5.25 inch drive, assembled from technologically inferior "off-the-shelf" components, was an "innovation" only in the sense that it was new. However, as this market grew and the drives improved, the companies that manufactured them eventually triumphed while many of the existing manufacturers of eight inch drives fell behind.</td></tr>
<tr><td>5.25 inch floppy disk drive</td><td>8 inch floppy disk drive</td></tr>
<tr><td>3.5 inch floppy disk drive</td><td>5.25 inch floppy disk drive</td></tr>
<tr><td>CDs and USB flash drives</td><td>Bernoulli drive and Zip drive</td></tr>
<tr><td>Display</td><td>Light-emitting diodes</td><td>Light bulbs</td><td>A LED is significantly smaller and less power-consuming than a light bulb. The first optical LEDs were weak, and only useful as indicator lights. Later models could be used for indoor lighting, and now several cities are switching to LED street lights. Incandescent light bulbs are being phased out in many countries. LED displays and AMOLED are also becoming competitive with LCDs.</td></tr>
</table>

	LCD LED displays	CRT	The first liquid crystal displays (LCD) were monochromatic and had low resolution. They were used in watches and other handheld devices, but during the early 2000s these (and other planar technologies) largely replaced the dominant cathode ray tube (CRT) technology for computer displays and television sets. CRT sets were very heavy, and the size and weight of the tube limited the maximum screen size to about 38 inches; in contrast, LCD and other flat-panel TVs are available in 40", 50", 60" and even bigger sizes, all of which weigh much less than a CRT set. CRT technologies did improve in the late 1990s with advances like true-flat panels and digital controls; however, these updates were not enough to prevent CRTs from being displaced by flat-panel LCD and LED TVs.
Manu-facturing	Hydraulic excavators	Cable-operated excavators	Hydraulic excavators were clearly innovative at the time of introduction but they gained widespread use only decades after. However, cable-operated excavators are still used in some cases, mainly for large excavations.
	Mini steel mills	Vertically integrated steel mills	By using mostly locally available scrap and power sources these mills can be cost effective even though not large.
	Plastic	Metal, wood, glass etc.	Bakelite and other early plastics had very limited use - their main advantages were electric insulation and low cost. New forms of plastic had advantages such as transparency, elasticity and combustibility. In the early 21st century, plastics can be used for many household items previously made of metal, wood and glass.
Medical	Ultrasound	Radiography (X-ray imaging)	Ultrasound technology is disruptive relative to X-ray imaging. Ultrasound was a new-market disruption. None of the X-ray companies participated in ultrasound until they acquired major ultrasound equipment companies.
Music	Digital synthesizer	Electronic organ, electric piano and piano	Synthesizers were initially low-cost, low-weight alternatives to electronic organs, electric pianos and acoustic pianos. In the 2010s, synthesizers are significantly cheaper than electric pianos and acoustic pianos, all while offering a much greater range of sound effects and musical sounds.
	Gramophone	Pianola	
	Downloadable Digital media	CDs, DVDs	In the 1990s, the music industry phased out the vinyl record single, leaving consumers with no means to purchase individual songs. This market was initially filled by illegal peer-to-peer file sharing technologies, and then by online retailers such as the iTunes Store and Amazon.com. This low end disruption eventually undermined the sales of physical, high-cost recordings such as records, tapes and CDs.

Photography	Digital photography	Chemical photography	Early digital cameras suffered from low picture quality and resolution and long shutter lag. Quality and resolution are no longer major issues in the 2010s and shutter lag issues have been largely resolved. The convenience of small memory cards and portable hard drives that hold hundreds or thousands of pictures, as well as the lack of the need to develop these pictures, also helped make digital cameras the market leader. Digital cameras have a high power consumption (but several lightweight battery packs can provide enough power for thousands of pictures). Cameras for classic photography are stand-alone devices. In the same manner, high-resolution digital video recording has replaced film stock, except for high-budget motion pictures and fine art.
	High speed CMOS video sensors	Photographic film	When first introduced, high speed CMOS sensors were less sensitive, had lower resolution, and cameras based on them had less duration (record time). The advantage of rapid setup time, editing in the camera, and nearly-instantaneous review quickly eliminated 16 mm high speed film systems. CMOS-based cameras also require less power (single phase 110 V AC and a few amps for CMOS, vs. 240 V single- or three-phase at 20-50 A for film cameras). Continuing advances have overtaken 35 mm film and are challenging 70 mm film applications.
Publishing	Computer printers	Offset printing	Offset printing has a high overhead cost, but very low unit cost compared to computer printers, and superior quality. But as printers, especially laser printers, have improved in speed and quality, they have become increasingly useful for creating documents in limited issues.
	Desktop publishing	Traditional publishing	Early desktop-publishing systems could not match high-end professional systems in either features or quality. Nevertheless, by the 2010s, they lowered the cost of entry to the publishing business, and economies of scale eventually enabled them to match, and then surpass, the functionality of the older dedicated publishing systems.
	Word Processing	Typewriter	The typewriter has been replaced with word processing software that has a wealth of functionality to stylize copy and facilitate document production.
Transportation	Steamboats	Sailing ships	The first steamships were deployed on inland waters where sailing ships were less effective, instead of on the higher profit margin seagoing routes. Hence steamships originally only competed in traditional shipping lines' "worst" markets.

	Automobiles	Rail transport	At the beginning of the 20th century, rail (including streetcars) was the fastest and most cost-efficient means of land transportation for goods and passengers in industrialized countries. The first cars, buses and trucks were used for local transportation in suburban areas, where they often replaced streetcars and industrial tracks. As highways expanded, medium- and later long-distance transports were relocated to road traffic, and some railways closed down. As rail traffic has a lower ton-kilometer cost, but a higher investment and operating cost than road traffic, rail is still preferred for large-scale bulk cargo (such as minerals). However, traffic congestion provides a bound on the efficiency of car use, and so rail is still used for urban passenger transport.
	high speed rail	short distance flights	In almost every market where high speed rail with journey times of two hours or less was introduced in competition with an air service, the air service was either greatly reduced within a few years or ceased entirely. Even in markets with longer rail travel times, airlines have reduced the amount of flights on offer and passenger numbers have gone down. Examples include the Barcelona-Madrid high speed railway, the Cologne Frankfurt high speed railway (where no direct flights are available as of 2016) or the Paris-London connection after the opening of High Speed 1
	Private jet	Supersonic transport	The Concorde aircraft has so far been the only supersonic airliner in extensive commercial traffic. However, it catered to a small customer segment, which could later afford small private sub-sonic jets. The loss of speed was compensated by flexibility and a more direct routing (i.e. no need to go through a hub). Supersonic flight is also banned above inhabited land, due to sonic booms. Concorde service ended in 2003.

Leapfrogging

The concept of leapfrogging is used in many different domains of economics and business, and was originally developed in the field of industrial organization and economic growth. The main idea beyond the concept of leapfrogging is that small and incremental innovations lead the dominant firm to stay ahead. However, sometimes, radical innovations will permit to new firms to leapfrog the ancient and dominant firm. The phenomenon can occur to firms but also to leadership of countries, or cities.

Leapfrogging in Industrial Organization

In the field of industrial organization (IO), the main work on leapfrogging was developed by Fudenberg, Gilbert, Stiglitz and TIrole (1983). In their article, they analyze under which conditions, a new entrant can leapfrog an established firm. Tirole (1988:391-2) has emphasized that leapfrogging can arise because an established monopolist has a somewhat reduced incentive to innovate because he is earning rents from the old technology. This is somehow based on Joseph Schumpeter's notion of 'gales of creative destruction'. The hypothesis proposes that companies holding monopolies based on incumbent technologies have less incentive to innovate than potential rivals, and therefore they eventually lose their technological leadership role when new radical technological innovations are adopted by new firms which are ready to take the risks. When the radical innovations eventually become the new technological paradigm, the newcomer companies leapfrog ahead of former leading firms.

Leapfrogging in Leadership and International Competition

In the field of economic growth, the question raised is: under which conditions, will a country which has the leadership lose its hegemony and be leapfrogged by another country. This has happened in history a few times. In the late eighteenth century, the Netherlands has been leapfrogged by the UK, which was the leader during the whole nineteenth century; this century was even coined Pax Britannica. After the second world war, it is the US which has leapfrogged the UK, and became the hegemonic power of the 20th century, Pax Americana.

Why is this happening? Brezis and Krugman (1993, 1997) suggest a mechanism that explains this pattern of "leapfrogging" as a response to occasional major changes in technology. In times of small and incremental technological change, increasing returns to scale tend to accentuate economic leadership. However, at times of a radical innovation and major technological breakthrough, economic leadership, since it also implies high wages, can deter the adoption of new ideas in the most advanced countries. A new technology may well seem initially inferior to older methods to those who have extensive experience with those older methods; yet that initially inferior technology may well have more potential for improvements and adaptation. When technological progress takes this form, economic leadership will tend to be the source of its own downfall.

In consequence, when a radical innovation occurs, it does not initially seem to be an improvement for leading nations, given their extensive experience with older technologies. Lagging nations have less experience; the new technique allows them to use their lower wages to enter the market. If the new technique proves more productive than the old, leapfrogging of leadership occurs.

Brezis and Krugman have applied this theory of leapfrogging to the field of geography, and explain why leading cities are often overtaken by upstart metropolitan areas. Such upheavals may be explained if the advantage of established urban centers rests on localized learning by doing. When a new technology is introduced, for which this accumulated experience is irrelevant, older centers prefer to stay with a technology in which they are more efficient. New centers, however, turn to the new technology and are competitive despite the raw state of that technology because of their lower land rents and wages. Over time, as the new technology matures, the established cities are overtaken.

Leapfrogging in Developing Countries

More recently the concept of leapfrogging is being used in the context of sustainable development for developing countries as a theory of development which may accelerate development by skipping inferior, less efficient, more expensive or more polluting technologies and industries and move directly to more advanced ones.[The mobile phone is also a wonderful example of a "leapfrog" technology: it has enabled developing countries to skip the fixed-line technology of the 20th century and move straight to the mobile technology of the 21st.] It is proposed that through leapfrogging developing countries can avoid environmentally harmful stages of development and do not need to follow the polluting development trajectory of industrialized countries. The adoption of solar energy technologies in developing countries are examples of where countries do not repeat the mistakes of highly industrialized countries in creating an energy infrastructure based on fossil fuels, but "jump" directly into the Solar Age.

Tunneling Through

A closely related concept is that of 'tunneling through' the Environmental Kuznets Curve (EKC). The concept proposes that developing countries could learn from the experiences of industrialized nations, and restructure growth and development to address potentially irreversible environmental damages from an early stage and thereby 'tunnel' through any prospective EKC. Environmental quality thereby does not have to get worse before it gets better and crossing safe limits or environmental thresholds can be avoided. Although in principle the concepts of leapfrogging (focused on jumping technological generations) and tunnelling through (focused on pollution) are distinct, in practice they tend to be conflated.

Millennium Development Goals

The concept of environmental leapfrogging also includes a social dimension. The diffusion and application of environmental technologies would not only reduce environmental impacts, but can at the same time contribute to sustainable economic development and the realization of the Millennium Development Goals (MDGs) by promoting greater access to resources and technologies to people who currently have no access. Regarding electricity currently nearly one third of the world population has no access to electricity and another third has only poor access. Reliance on traditional biomass fuels for cooking and heating can have a serious impact on health and the environment. There is not only a direct positive link between sustainable renewable energy technologies and climate change mitigation, but also between clean energy and issues of health, education and gender equity.

Examples

A frequently cited example is countries which move directly from having no telephones to having cellular phones, skipping the stage of copperwire landline telephones altogether.

The use of ethanol fuel in Brazil, where ethanol produced from sugarcane for transportation replaces gasoline, provides evidence that leapfrogging is a possible alternative to business-as-usual development.

The city of Rizhao in China is an example of energy leapfrogging on city level. As of 2007, 99 percent of households in the central districts use solar water heaters, and most of the lighting and traffic signals are powered with photovoltaic solar power.

Necessary Conditions

Leapfrogging can occur accidentally, when the only systems around for adoption are better than legacy systems elsewhere, or situationally, such as the adoption of decentralized communication for a sprawling, rural countryside. It may also be initiated intentionally, e.g. by policies promoting the installation of WiFi and free computers in poor urban areas.

The Reut Institute has carried out extensive research regarding the common denominators of all the different countries that have successfully 'leapt' in recent years. It concludes that to leapfrog a country needs to create a shared vision, leadership by a committed elite, 'Inclusive growth', relevant institutions, a labor market suited to cope with rapid growth and changes, growth diagnostics of the country's bottlenecks and focused reforms as well as local and regional development and national mobilization.

Promotion by International Initiatives

Japan's Low-Carbon Society 2050 Initiative has the objective to cooperate with and offer support to Asian developing countries to leapfrog towards a low-carbon energy future.

Paradigm Shift

A paradigm shift, as identified by American physicist and philosopher Thomas Kuhn, is a fundamental change in the basic concepts and experimental practices of a scientific discipline. Kuhn contrasted these shifts, which characterize a scientific revolution, to the activity of normal science, which he described as scientific work done within a prevailing framework (or paradigm).

The nature of scientific revolutions has been a question posed by modern philosophy since Immanuel Kant used the phrase in the preface to his *Critique of Pure Reason* (1781), referring to Greek mathematics and Newtonian physics. In the 20th century, new crises in the basic concepts of mathematics, physics, and biology, revitalized interest in the question among scholars. It was against this active background that Kuhn published his work.

Kuhn presented his notion of a paradigm shift in his influential book *The Structure of Scientific Revolutions* (1962). As one commentator summarizes:

Kuhn acknowledges having used the term "paradigm" in two different meanings. In the first one, "paradigm" designates what the members of a certain scientific community have in common, that is to say, the whole of techniques, patents and values shared by the members of the community. In the second sense, the paradigm is a single element of a whole, say for instance Newton's Principia, which, acting as a common model or an example... stands for the explicit rules and thus defines a coherent tradition of investigation. Thus the question is for Kuhn to investigate by means of the paradigm what makes possible the constitution of what he calls "normal science". That is to say, the science which can decide if a certain problem will be considered scientific or not. Normal science does not mean at all a science guided by a coherent system of rules, on the contrary, the rules can be derived from the paradigms, but the paradigms can guide the investigation also in the absence of rules. This is precisely the second meaning of the term "paradigm", which Kuhn considered the most new and profound, though it is in truth the oldest.

Since the 1960s, the concept of a paradigm shift has also been used in numerous non-scientific contexts to describe a profound change in a fundamental model or perception of events, even though Kuhn himself restricted the use of the term to the hard sciences.

In a 2015 retrospective on Kuhn, the philosopher Martin Cohen describes the notion of the 'Paradigm Shift' as a kind of intellectual virus – spreading from hard science to social science and on to the arts and even everyday political rhetoric today. Cohen claims that Thomas Kuhn himself had only a very hazy idea of what it might mean and, in line with the American philosopher of science, Paul Feyerabend, accuses Kuhn of retreating from the more radical implications of his theory, which are that scientific facts are never really more than opinions, whose popularity is transitory and far from conclusive.

Kuhnian Paradigm Shifts

An epistemological paradigm shift was called a "scientific revolution" by epistemologist and historian of science Thomas Kuhn in his book *The Structure of Scientific Revolutions*.

Kuhn used the duck-rabbit optical illusion to demonstrate the way in which a paradigm shift could cause one to see the same information in an entirely different way.

A scientific revolution occurs, according to Kuhn, when scientists encounter anomalies that cannot be explained by the universally accepted paradigm within which scientific progress has thereto been made. The paradigm, in Kuhn's view, is not simply the current theory, but the entire worldview in which it exists, and all of the implications which come with it. This is based on features of landscape of knowledge that scientists can identify around them.

There are anomalies for all paradigms, Kuhn maintained, that are brushed away as acceptable levels of error, or simply ignored and not dealt with (a principal argument Kuhn uses to reject Karl Popper's model of falsifiability as the key force involved in scientific change). Rather, according to Kuhn, anomalies have various levels of significance to the practitioners of science at the time. To put it in the context of early 20th century physics, some scientists found the problems with calculating Mercury's perihelion more troubling than the Michelson-Morley experiment results, and some the other way around. Kuhn's model of scientific change differs here, and in many places, from that of the logical positivists in that it puts an enhanced emphasis on the individual humans involved as scientists, rather than abstracting science into a purely logical or philosophical venture.

When enough significant anomalies have accrued against a current paradigm, the scientific discipline is thrown into a state of *crisis,* according to Kuhn. During this crisis, new ideas, perhaps ones previously discarded, are tried. Eventually a *new* paradigm is formed, which gains its own new followers, and an intellectual "battle" takes place between the followers of the new paradigm and the hold-outs of the old paradigm. Again, for early 20th century physics, the transition between the Maxwellian electromagnetic worldview and the Einsteinian relativistic worldview was neither instantaneous nor calm, and instead involved a protracted set of "attacks," both with empirical data as well as rhetorical or philosophical arguments, by both sides, with the Einsteinian theory winning out in the long run. Again, the weighing of evidence and importance of new data was fit through the human sieve: some scientists found the simplicity of

Einstein's equations to be most compelling, while some found them more complicated than the notion of Maxwell's aether which they banished. Some found Eddington's photographs of light bending around the sun to be compelling, while some questioned their accuracy and meaning. Sometimes the convincing force is just time itself and the human toll it takes, Kuhn said, using a quote from Max Planck: "a new scientific truth does not triumph by convincing its opponents and making them see the light, but rather because its opponents eventually die, and a new generation grows up that is familiar with it."

After a given discipline has changed from one paradigm to another, this is called, in Kuhn's terminology, a *scientific revolution* or a *paradigm shift*. It is often this final conclusion, the result of the long process, that is meant when the term *paradigm shift* is used colloquially: simply the (often radical) change of worldview, without reference to the specificities of Kuhn's historical argument.

Science and Paradigm Shift

A common misinterpretation of paradigms is the belief that the discovery of paradigm shifts and the dynamic nature of science (with its many opportunities for subjective judgments by scientists) are a case for relativism: the view that all kinds of belief systems are equal. Kuhn vehemently denies this interpretation and states that when a scientific paradigm is replaced by a new one, albeit through a complex social process, the new one is *always better*, not just different.

These claims of relativism are, however, tied to another claim that Kuhn does at least somewhat endorse: that the language and theories of different paradigms cannot be translated into one another or rationally evaluated against one another — that they are *incommensurable*. This gave rise to much talk of different peoples and cultures having radically different worldviews or conceptual schemes — so different that whether or not one was better, they could not be understood by one another. However, the philosopher Donald Davidson published a highly regarded essay in 1974, "*On the Very Idea of a Conceptual Scheme*" (*Proceedings and Addresses of the American Philosophical Association*, Vol. 47, (1973-1974), pp. 5–20) arguing that the notion that any languages or theories could be incommensurable with one another was itself incoherent. If this is correct, Kuhn's claims must be taken in a weaker sense than they often are. Furthermore, the hold of the Kuhnian analysis on social science has long been tenuous with the wide application of multi-paradigmatic approaches in order to understand complex human behaviour.

Paradigm shifts tend to be most dramatic in sciences that appear to be stable and mature, as in physics at the end of the 19th century. At that time, physics seemed to be a discipline filling in the last few details of a largely worked-out system. In 1900, Lord Kelvin famously told an assemblage of physicists at the British Association for the Advancement of Science, "There is nothing new to be discovered in physics now. All that remains is more and more precise measurement." Five years later, Albert Einstein published his paper on special relativity, which challenged the very simple set of rules laid down by Newtonian mechanics, which had been used to describe force and motion for over two hundred years.

In *The Structure of Scientific Revolutions*, Kuhn wrote, "Successive transition from one paradigm to another via revolution is the usual developmental pattern of mature science." (p. 12) Kuhn's idea was itself revolutionary in its time, as it caused a major change in the way that academics talk about science. Thus, it could be argued that it caused or was itself part of a "paradigm shift" in the history and sociology of science. However, Kuhn would not recognise such a paradigm shift. In the social sciences, people can still use earlier ideas to discuss the history of science.

Philosophers and historians of science, including Kuhn himself, ultimately accepted a modified version of Kuhn's model, which synthesizes his original view with the gradualist model that preceded it.

Examples of Paradigm Shifts

Natural Sciences

Some of the "classical cases" of Kuhnian paradigm shifts in science are:

- 1543 - The transition in cosmology from a Ptolemaic cosmology to a Copernican one.
- The transition in optics from geometrical optics to physical optics.
- 1543 - The acceptance of the work of Andreas Vesalius, whose work *De humani corporis fabrica* corrected the numerous errors in the previously-held system created by Galen.
- 1687 - The transition in mechanics from Aristotelian mechanics to classical mechanics.
- 1783 - The acceptance of Lavoisier's theory of chemical reactions and combustion in place of phlogiston theory, known as the chemical revolution.
- 1826 - The discovery of hyperbolic geometry.
- 1866 - The acceptance of Mendelian inheritance, as opposed to pangenesis in the early 20th century.
- The acceptance of the theory of biogenesis, that all life comes from life, as opposed to the theory of spontaneous generation, which began in the 17th century

and was not complete until the 19th century with Louis Pasteur.

- 1905 - The development of quantum mechanics, which replaced classical mechanics at microscopic scales.
- 1905 - The transition from the luminiferous aether present in space to electromagnetic radiation in spacetime.
- 1920 - The transition between the worldview of Newtonian physics and the Einsteinian relativistic worldview.
- The development of absolute dating.
- 1965 - The acceptance of plate tectonics as the explanation for large-scale geologic changes.

Social Sciences

In Kuhn's view, the existence of a single reigning paradigm is characteristic of the natural sciences, while philosophy and much of social science were characterized by a "tradition of claims, counterclaims, and debates over fundamentals." Others have applied Kuhn's concept of paradigm shift to the social sciences.

- The movement known as the cognitive revolution moved away from behaviourist approaches to psychological study and the acceptance of cognition as central to studying human behaviour.
- The Keynesian revolution is typically viewed as a major shift in macroeconomics. According to John Kenneth Galbraith, Say's Law dominated economic thought prior to Keynes for over a century, and the shift to Keynesianism was difficult. Economists who contradicted the law, which implied that underemployment and underinvestment (coupled with oversaving) were virtually impossible, risked losing their careers. In his magnum opus, Keynes cited one of his predecessors, John A. Hobson, who was repeatedly denied positions at universities for his heretical theory.
- Later, the movement for monetarism over Keynesianism marked a second divisive shift. Monetarists held that fiscal policy was not effective for stabilizing inflation, that it was solely a monetary phenomenon, in contrast to the Keynesian view of the time was that both fiscal and monetary policy were important. Keynesians later adopted much of the monetarists' view of the quantity theory of money and shifting Phillips curve, theories they initially rejected.
- First proposed by Ferdinand de Saussure in 1879, the laryngeal theory in Indo-European linguistics postulated the existence of "laryngeal" consonants in the Proto-Indo-European language (PIE), a theory that was confirmed by the discovery of the Hittite language in the early 20th century. The theory has since

been accepted by the vast majority of linguists, paving the way for the internal reconstruction of the syntax and grammatical rules of PIE and is considered one of the most significant developments in linguistics since the initial discovery of the Indo-European language family.

Marketing

In the later part of the 1990s, 'paradigm shift' emerged as a buzzword, popularized as marketing speak and appearing more frequently in print and publication. In his book *Mind The Gaffe*, author Larry Trask advises readers to refrain from using it, and to use caution when reading anything that contains the phrase. It is referred to in several articles and books as abused and overused to the point of becoming meaningless.

Other Uses

The term "paradigm shift" has found uses in other contexts, representing the notion of a major change in a certain thought-pattern — a radical change in personal beliefs, complex systems or organizations, replacing the former way of thinking or organizing with a radically different way of thinking or organizing:

- M. L. Handa, a professor of sociology in education at O.I.S.E. University of Toronto, Canada, developed the concept of a paradigm within the context of social sciences. He defines what he means by "paradigm" and introduces the idea of a "social paradigm". In addition, he identifies the basic component of any social paradigm. Like Kuhn, he addresses the issue of changing paradigms, the process popularly known as "paradigm shift." In this respect, he focuses on the social circumstances which precipitate such a shift. Relatedly, he addresses how that shift affects social institutions, including the institution of education.

- The concept has been developed for technology and economics in the identification of new techno-economic paradigms as changes in technological systems that have a major influence on the behaviour of the entire economy (Carlota Perez; earlier work only on technological paradigms by Giovanni Dosi). This concept is linked to Joseph Schumpeter's idea of creative destruction. Examples include the move to mass production and the introduction of microelectronics.

- Two photographs of the Earth from space, "Earthrise" (1968) and "The Blue Marble" (1972), are thought to have helped to usher in the environmentalist movement which gained great prominence in the years immediately following distribution of those images.

- Hans Küng applies Thomas Kuhn's theory of paradigm change to the entire history of Christian thought and theology. He identifies six historical "macromodels": 1) the apocalyptic paradigm of primitive Christianity, 2) the Hellenistic paradigm of the patristic period, 3) the medieval Roman Catholic paradigm, 4) the Protestant (Reformation) paradigm, 5) the modern Enlightenment paradigm, and 6) the emerging ecumenical paradigm. He also discusses five analogies between natural science and theology in relation to paradigm shifts. Küng addresses paradigm change in his books, Paradigm Change in Theology and *Theology for the Third Millennium: An Ecumenical View.*

Eureka Effect

The eureka effect (also known as the aha! moment or eureka moment) refers to the common human experience of suddenly understanding a previously incomprehensible problem or concept. Some research describes the aha! effect (also known as insight or epiphany) as a memory advantage, but conflicting results exist as to where exactly it occurs in the brain, and it is difficult to predict under what circumstances one can predict an Aha! moment.

A 16th century woodcut of Archimedes' eureka moment

Insight is a psychological term that attempts to describe the process in problem solving when a previously unsolvable puzzle becomes suddenly clear and obvious. Often this transition from not understanding to spontaneous comprehension is accompanied by an exclamation of joy or satisfaction, an Aha! moment. A person utilizing insight to solve a problem is able to give accurate, discrete, all-or-nothing type responses, whereas individuals not using the insight process are more likely to produce partial, incomplete responses.

A recent theoretical account of the Aha! moment started with four defining attributes of this experience. First, the Aha! moment appears suddenly; second, the solution to a problem can be processed smoothly, or fluently; third, the Aha! moment elicits positive affect; fourth, a person experiencing the Aha! moment is convinced that a solution is true. These four attributes are not separate but can be combined because the experience of processing fluency, especially when it occurs surprisingly (for example, because it is sudden), elicits both positive affect and judged truth.

Insight can be conceptualized as a two phase process. The first phase of an Aha! experience requires the problem solver to come upon an impasse, where they become stuck and even though they may seemingly have explored all the possibilities, are still unable to retrieve or generate a solution. The second phase occurs suddenly and unexpectedly. After a break in mental fixation or re-evaluating the problem, the answer is retrieved. Some research suggest that insight problems are difficult to solve because of our mental fixation on the inappropriate aspects of the problem content. In order to solve insight problems, one must "think outside the box". It is this elaborate rehearsal that may cause people to have better memory for Aha! moments. Insight is believed to occur with a break in mental fixation, allowing the solution to appear transparent and obvious.

History and Etymology

The effect is named from a story about the ancient Greek polymath Archimedes. In the story, Archimedes was asked (c. 250 BC) by the local king to determine whether a crown was pure gold. During a subsequent trip to a public bath, Archimedes noted that water was displaced when his body sank into the bath, and particularly that the volume of water displaced equaled the volume of his body immersed in the water. Having discovered how to measure the volume of an irregular object, and conceiving of a method to solve the king's problem, Archimedes allegedly leaped out and ran home naked, shouting "eureka" (I have found it). This story is now thought to be fictional, because it was first mentioned by the Roman writer Vitruvius nearly 200 years after the date of the alleged event, and because the method described by Vitruvius would not have worked. However, Archimedes certainly did important, original work in hydrostatics, notably in his *On Floating Bodies*.

Research

Initial Research

The Eureka effect was first described by Pamela Auble, Jeffrey Franks and Salvatore Soraci in 1979. The subject would be presented with an initially confusing sentence such as "The haystack was important because the cloth ripped". After a certain period of time of non-comprehension by the reader, the cue word (parachute) would be presented, the reader could comprehend the sentence, and this resulted in better recall on memory tests. Subjects spend a considerable amount of time attempting to solve the problem, and initially it was hypothesized that elaboration towards comprehension may play a role in increased recall. There was no evidence that elaboration had any effect for recall. It was found that both "easy" and "hard" sentences that resulted in an Aha! effect had significantly better recall rates than sentences that subjects were able to comprehend immediately. In fact equal recall rates were obtained for both "easy" and "hard" sentences which were initially noncomprehensible. It seems to be this noncomprehension to comprehension which results in better recall. The essence of the aha feeling underling insight problem solving was systemically empirically investigated by Shen and his colleagues.

How People Solve Insight Problems

Currently there are two theories for how people arrive at the solution for insight problems. The first is the progress monitoring theory. The person will analyze the distance from their current state to the goal state. Once a person realizes that they cannot solve the problem while on their current path, they will seek alternative solutions. In insight problems this usually occurs late in the puzzle. The second way that people attempt to solve these puzzles is the representational change theory. The problem solver initially has a low probability for success because they use inappropriate knowledge as they set unnecessary constraints on the problem. Once the person relaxes his or her constraints, they can bring previously unavailable knowledge into working memory to solve the problem. The person also utilizes chunk decomposition, where he or she will separate meaningful chunks into their component pieces. Both constraint relaxation and chunk decomposition allow for a change in representation, that is, a change in the distribution of activation across working memory, at which point they may exclaim "aha!". Currently both theories have support, with the progress monitoring theory being more suited to multiple step problems, and the representational change theory more suited to single step problems.

The Eureka effect on memory occurs only when there is an initial confusion. When subjects were presented with a clue word before the confusing sentence was presented, there was no effect on recall. If the clue was provided after the sentence was presented, an increase in recall occurred.

Memory

It had been determined that recall is greater for items that were generated by the subject versus if the subject was presented with the stimuli. There seems to be a memory advantage for instances where people are able to produce an answer themselves, recall was higher when Aha! reactions occurred. They tested sentences that were initially hard to understand, but when presented with a cued word, the comprehension became more apparent. Other evidence was found indicating that effort in processing visual stimuli was recalled more frequently than the stimuli that were simply presented. This study was done using connect-the-dots or verbal instruction to produce either a nonsense or real image. It is believed that effort made to comprehend something when encoding induces activation of alternative cues that later participate in recall.

Cerebral lateralization

Functional magnetic resonance imaging and electroencephalogram studies have found that problem solving requiring insight involves increased activity in the right cerebral hemisphere as compared with problem solving not requiring insight. In particular, increased activity was found in the right hemisphere anterior superior temporal gyrus.

Sleep

Some unconscious processing may take place while a person is asleep, and there are several cases of scientific discoveries coming to people in their dreams. Friedrich August Kekulé von Stradonitz said that the ring structure of benzene came to him in a dream where a snake was eating its own tail. Studies have shown increased performance at insight problems if the subjects slept during a break between receiving the problem and solving it. Sleep may function to restructure problems, and allow new insights to be reached. Henri Poincaré stated that he valued sleep as a time for "unconscious thought" that helped him break through problems.

Other Theories

Professor Stellan Ohlsson believes that at the beginning of the problem-solving process, some salient features of the problem are incorporated into a mental representation of the problem. In the first step of solving the problem, it is considered in the light of previous experience. Eventually, an impasse is reached, where all approaches to the problem have failed, and the person becomes frustrated. Ohlsson believes that this impasse drives unconscious processes which change the mental representation of a problem, and cause novel solutions to occur.

General Procedure for Conducting ERP and EEG Studies

When studying insight, or the Aha! Effect, ERP or EEG general methods are used. Initially a baseline measurement is taken, which generally asks the subject to simply remember an answer to a question. Following this, subjects are asked to focus on the screen while a logogriph is shown, and then they are given time with a blank screen to get the answer, once they do they are required to press a key. After which the answer appears on the screen. The subjects are then asked to press one key to indicate that they thought of the correct answer and another to indicate if they got the answer wrong, finally, not to press a key at all if they were unsure or did not know the answer.

Evidence in EEG Studies

Resting-state neural activity has a standing influence on cognitive strategies used when solving problems, particularly in the case of deriving solutions by methodical search or by sudden insight. The two cognitive strategies used involve both search and analysis of current state of a problem, to the goal state of that problem, while insight problems are a sudden awareness of the solution to a problem.

Subjects studied were first recorded on the base-line resting state of thinking. After being tested using the method described in the General Procedure for Conducting ERP and EEG Studies, the ratio of insight versus non-insight solution were made to determine whether an individual is classified as a high insight (HI) or a low insight (LI) individual. Discriminating between HI and LI individuals were important as both groups use different cognitive strategies to solve anagram problems used in this study. Right hemisphere activation is believed to be involved in Aha! effects, so it comes as no surprise that HI individuals would show greater activation in the right hemisphere than the left hemisphere when compared to the LI individuals. Evidence was found to support this idea, there was greater activation in HI subjects at the right dorsal-frontal (low-alpha band), right inferior-frontal (beta and gamma bands) and the right parietal (gamma band) areas. As for LI subjects, left inferior-frontal and left anterior-temporal areas were active (low-alpha band).

There were also differences in attention between individuals of HI and LI. It has been suggested that individuals who are highly creative exhibit diffuse attention, thus allowing them a greater range of environmental stimuli. It was found that individuals who displayed HI would have less resting state occipital alpha-band activity, meaning there would be less inhibition of the visual system. Individuals that were less creative were found to focus their attention, thus causing them to sample less of their environment. Although, LI individuals were shown to have more occipital beta activity, consistent with heightened focused attention.

Evidence in ERP Studies

These results are more reflective of models, rather than empirical evidence, as source localization is hard to determine precisely. Due to the nature of these studies that use Chinese logographs, there is a difficulty in an exact translation; a language barrier certainly exists.

There are some difficulties that exist in brain imaging when it comes to insight, thus making it hard to discuss neural mechanisms. Issues include: that insight occurs when an unwarranted mental fixation is broken and when novel task related associations are formed on top of old cognitive skills.

One theory discussed found that “Aha” answers produced more negative ERP results, N380 in the ACC, than the “No-Aha” answers, 250–500 ms, after an answer was produced. The authors suspected that this N380 in the ACC, which plays the role of a warning sign of breaking the mental set, was a reflection of the Aha! effect. Another study was done showed that an Aha! effect was elicited at N320 which has a strong activation in the central-posterior region. These previous studies reflective the premise of the study, that the Aha! effect occurs in the anterior cingulate cortex, while this study finds results indicating the posterior cingulate cortex is responsible. It was found that there was a N350 in the posterior cingulate cortex for successful guessing of logographs, not in the anterior cingulate cortex. The posterior cingulate cortex seems to play a more non-executive function in monitoring and inhibiting the mind set and cognitive function.

Another significant finding of this study done by Qiu and Zhang (2008), was a late positive component (LPC) in successful guessing of the logograph and then recognition of the answer at 600 and 700 ms, post-stimulus, in the parahippocampal gyrus (BA34). The data suggests that the parahippocampus is involved in searching of a correct answer by manipulating it in working memory, and integrating relationships between the base of the target logograph. The parahippocampal gyrus may reflect the formation of novel associations while solving insight problem.

Another ERP study is fairly similar to the Qiu and Zhang, 2008 study, however, this study claims to have anterior cingulate cortex activation at N380, which may be responsible for the mediation of breaking the mental set. Other areas of interest were prefrontal cortex (PFC), the posterior parietal cortex, and the medial temporal lobe. If subjects failed to solve the riddle, and then were shown the correct answer, they displayed the feeling of insight, which reflected the electroencephalogram recordings.

Overall, it is quite apparent that there are many aspects that can explain the Aha! effect. No particular area has been determined but from the information gathered, it seems that insight occurs in many parts of the brain, within a given time period.

Evidence in fMRI studies

A study with the goal of recording the activity that occurs in the brain during an Aha! moment using fMRIs was conducted in 2003 by Jing Luo and Kazuhisa Niki. Participants in this study were presented with a series of Japanese riddles, and asked to rate their impressions toward each question using the following scale: (1) I can understand this question very well and know the answer; (2) I can understand this question very well and feel it is interesting, but I do not know the answer; or (3) I cannot understand this question and do not know the answer. This scale allowed the researchers to only look at participants who would experience an Aha! moment upon viewing the answer to the riddle. In previous studies on insight, researchers have found that participants reported feelings of insight when they viewed the answer to an unsolved riddle or problem. Luo and Niki had the goal of recording these feelings of insight in their participants using fMRIs. This method allowed the researchers to directly observe the activity that was occurring in the participant's brains during an Aha! moment.

> An example of a Japanese riddle used in the study: *The thing that can move heavy logs, but cannot move a small nail → A river.*

Participants were given 3 minutes to respond to each riddle, before the answer to the riddle was revealed. If the participant experienced an Aha! moment upon viewing the correct answer, any brain activity would be recorded on the fMRI. The fMRI results for this study showed that when participants were given the answer to an unsolved riddle, the activity in their right hippocampus increased significantly during these Aha! moments. This increased activity in the right hippocampus may be attributed to the formation of new associations between old nodes. These new associations will in turn strengthen memory for the riddles and their solutions.

Although various studies using EEGs, ERPs, and fMRI's report activation in a variety of areas in the brain during Aha! moments, it is interesting to note that this activity occurs predominantly in the right hemisphere.

Insight Problems and Problems with Insight

The Nine Dot Problem

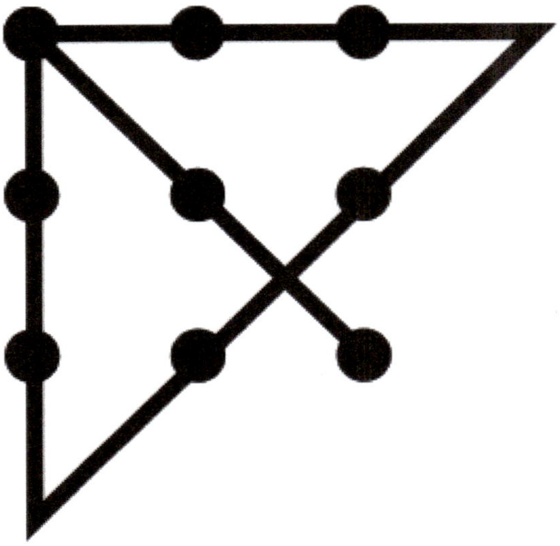

The Nine Dot Problem with solution. Most individuals fail to draw lines beyond the dots that compose the square, and are unable to solve this puzzle.

The Nine Dot Problem is a classic spatial problem used by psychologists to study insight. The problem consists of 3 × 3 square created by 9 black dots. The task is to connect all 9 dots using exactly 4 straight lines, without retracing or removing one's pen from the paper. Kershaw & Ohlsson report that in a laboratory setting with a time limit of 2 or 3 minutes, the expected solution rate is 0%.

The difficulty with the Nine Dot Problem is that it requires respondents to look beyond the conventional figure-ground relationships that create subtle, illusory spatial constraints and (literally) "think outside of the box". Breaking the spatial constraints shows a shift in attention in working memory and utilizing new knowledge factors to solve the puzzle.

Verbal Riddles

Verbal riddles are becoming popular problems in insight research.

Example: "A man was washing windows on a high-rise building when he fell from the 40-foot ladder to the concrete path below. Amazingly, he was unhurt. Why? [Answer] He slipped from the bottom rung!"

Matchstick Arithmetic

Matchstick arithmetic, which was developed and used by G. Knoblich, involves matchsticks that are arranged to show a simple but incorrect math equation in Roman numerals. The task is to correct the equation by moving only one matchstick.

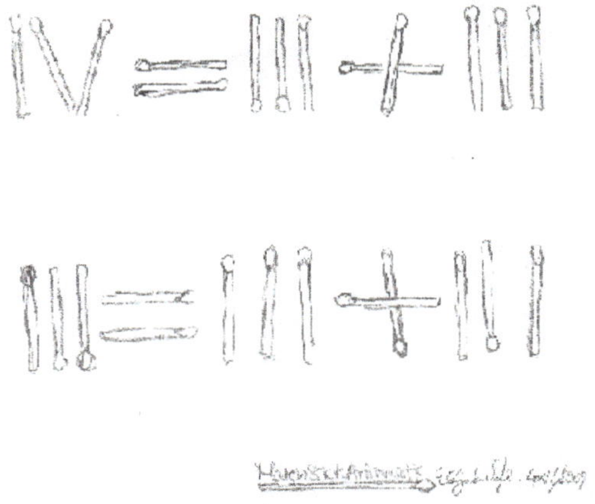

Two examples of Matchstick Arithmetic Problems.

Anagrams

Anagrams involve manipulating the order of a given set of letters in order to create one or many words. The original set of letters may be a word itself, or simply a jumble.

Example: *Santa* can be transformed to spell *Satan*.

Rebus Puzzles

Rebus puzzles, also called "wordies", involve verbal and visual cues that force the respondent to restructure and "read between the lines" (almost literally) to solve the puzzle.

Some examples:

you just me [answer: just between you and me]

PUNISHMENT [answer: capital punishment]

i i i i i i i i
oooooooo [answer: circles under the eyes]

Remote Associates Test (RAT)

The Remote Associates Test (known as the RAT) was developed by Martha Mednick in 1962 to test creativity. However, it has recently been utilized in insight research.

The test consists of presenting participants with a set of words, such as *lick*, *mine*, and *shaker*. The task is to identify the word that connects these three seemingly unrelated ones. In this example, the answer is *salt*. The link between words is associative, and does not follow rules of logic, concept formation or problem solving, and thus requires the respondent to work outside of these common heuristical constraints.

Performance on the RAT is known to correlate with performance on other standard insight problems.

The Eight Coin Problem

In this problem a set of 8 coins is arranged on a table in a certain configuration, and the subject is told to move 2 coins so that all coins touch exactly three others. The difficulty in this problem comes from thinking of the problem in a purely 2-dimensional way, when a 3-dimensional approach is the only way to solve the problem.

Problems with Insight

Insight research is problematic because of the ambiguity and lack of agreement among psychologists of its definition. This could largely be explained by the phenomenological nature of insight, and the difficulty in catalyzing its occurrence, as well as the ways in which it is experimentally "triggered".

The pool of insight problems currently employed by psychologists is small and tepid, and due to its heterogeneity and often high difficulty level, is not conducive of validity or reliability.

One of the biggest issues surrounding insight problems is that for most participants, they're simply too difficult. For many problems, this difficulty revolves around the requisite restructuring or re-conceptualization of the problem or possible solutions, for example, drawing lines beyond the square composed of dots in the Nine-Dot Problem.

Furthermore, there are issues related to the taxonomy of insight problems. Puzzles and problems that are utilized in experiments to elicit insight may be classified in two ways. "Pure" insight problems are those that necessitate the use of insight, whereas "hybrid" insight problems are those that can be solved by other methods, such as the trial and error. As Weisberg (1996) points out, the existence of hybrid problems in insight research poses a significant threat to any evidence gleaned from studies that employ them. While the phenomenological experience of insight can help to differentiate insight-solving from non-insight solving (by asking the respondent to describe how they solved the problem, for example), the risk that non-insight solving has been mistaken for insight solving still exists. Likewise, issues surrounding the validity of insight evidence is also threatened by the characteristically small sample sizes. Experimenters may recruit an initially adequate sample size, but because of the level of difficulty inherent to insight problems, only a small fraction of any sample will successfully solve the puzzle or task given to them; placing serious limits on usable data. In the case of studies using hybrid problems, the final sample is at even greater risk of being very small by way of having to exclude whatever percentage of respondents solved their given puzzle without utilizing insight.

The Aha! Effect and Scientific Discovery

There are several examples of scientific discoveries being made due to a sudden flash of insight. One of the key insights in developing his special theory of relativity came to Albert Einstein while talking to his friend Michele Besso:

I started the conversation with him in the following way: "Recently I have been working on a difficult problem. Today I come here to battle against that problem with you." We discussed every aspect of this problem. Then suddenly I understood where the key to this problem lay. Next day I came back to him again and said to him, without even saying hello, "Thank you. I've completely solved the problem."

However, Einstein has said that the whole idea of special relativity did not come to him as a sudden, single eureka moment, and that he was "led to it by steps arising from the individual laws derived from experience". Similarly, Carl Friedrich Gauss said after a eureka moment: "I have the result, only I do not yet know how to get to it."

Sir Alec Jeffreys had a eureka moment in his lab in Leicester after looking at the X-ray film image of a DNA experiment at 9:05 am on Monday 10 September 1984, which unexpectedly showed both similarities and differences between the DNA of different members of his technician's family. Within about half an hour, he realized the possible scope of DNA fingerprinting, which uses variations in the genetic code to identify individuals. The method has become important in forensic science to assist police detective work, and it has also proved useful in resolving paternity and immigration disputes. The method can also be applied to non-human species, for example in wildlife population genetics studies. Before his methods were commercialised in 1987, his laboratory was the only centre carrying out DNA fingerprinting in the world.

In popular Psychology

Oprah Winfrey

Popular culture has its own views of the Aha! effect. Although both are defined as that moment of insight that changes one's state of mind, its applications greatly differ. Where scientists have focused on understanding the mechanisms of insights as well as how and where Aha! moments occur in the brain, Oprah Winfrey has taken this phenomenon and turned it into a popular and well-recognized state of mind. By focusing on the emotional and life changing occurrences an individual experiences, Oprah views the Aha! effect as a sudden moment of realization, where an individual recognizes that they need to make a change or move forward in life. Oprah includes Aha! Moments in her monthly magazine, *O: The Oprah Magazine*, and commonly makes reference to these moments of insight on her television show, *Oprah*.

Creative Destruction

Crowd at New York's American Union Bank during a bank run early in the Great Depression. Marx argued that the devaluation of wealth during capitalism's periodic financial crises was an inevitable outcome of the processes of wealth creation.

Creative destruction (German: *schöpferische Zerstörung*), sometimes known as Schumpeter's gale, is a concept in economics which since the 1950s has become most readily identified with the Austrian American economist Joseph Schumpeter who derived it from the work of Karl Marx and popularized it as a theory of economic innovation and the business cycle.

According to Schumpeter, the "gale of creative destruction" describes the "process of industrial mutation that incessantly revolutionizes the economic structure from within, incessantly destroying the old one, incessantly creating a new one". In Marxist economic theory the concept refers more broadly to the linked processes of the accumulation and annihilation of wealth under capitalism.

The German Marxist sociologist Werner Sombart has been credited with the first use of these terms in his work *Krieg und Kapitalismus* ("War and Capitalism", 1913). In the earlier work of Marx, however, the idea of creative destruction or annihilation (German: *Vernichtung*) implies not only that capitalism destroys and reconfigures previous economic orders, but also that it must ceaselessly devalue existing wealth (whether through war, dereliction, or regular and periodic economic crises) in order to clear the ground for the creation of new wealth.

In *Capitalism, Socialism and Democracy* (1942), Joseph Schumpeter developed the concept out of a careful reading of Marx's thought (to which the whole of Part I of the book is devoted), arguing (in Part II) that the creative-destructive forces unleashed by capitalism would eventually lead to its demise as a system. Despite this, the term subsequently gained popularity within neoliberal or free-market economics as a description of processes such as downsizing in order to increase the efficiency and dynamism of a company. The Marxian usage has, however, been retained and further developed in the work of social scientists such as David Harvey, Marshall Berman, Manuel Castells and Daniele Archibugi.

History

In Marx's Thought

Although the modern term "creative destruction" is not used explicitly by Marx, it is largely derived from his analyses, particularly in the work of Werner Sombart (whom Engels described as the only German professor who understood Marx's *Capital*), and of Joseph Schumpeter, who discussed at length the origin of the idea in Marx's work.

In *The Communist Manifesto* of 1848, Karl Marx and Friedrich Engels described the crisis tendencies of capitalism in terms of "the enforced destruction of a mass of productive forces":

> Modern bourgeois society, with its relations of production, of exchange and of property, a society that has conjured up such gigantic means of production and of exchange, is like the sorcerer who is no longer able to control the powers of the nether world whom he has called up by his spells. [...] It is enough to mention the commercial crises that by their periodical return put the existence of the whole of bourgeois society on trial, each time more threateningly. In these crises, *a great part not only of existing production, but also of previously created productive forces, are periodically destroyed*. In these crises, there breaks out an epidemic that, in all earlier epochs, would have seemed an absurdity — the epidemic of over-production. Society suddenly finds itself put back into a state of momentary barbarism; it appears as if a famine, a universal war of devastation, had cut off the supply of every means of subsistence; industry and commerce seem to be destroyed; and why? Because there is too much civilisation, too much means of subsistence, too much industry, too much commerce. The productive forces at the disposal of society no longer tend to further the development of the conditions of bourgeois property; on the contrary, they have become too powerful for these conditions. [...] And how does the bourgeoisie get over these crises? On the one hand by *enforced destruction of a mass of productive forces*; on the other, by the conquest of new markets, and by the more thorough exploitation of the old ones. That is to say, by paving the way for more

> extensive and more destructive crises, and by diminishing the means whereby crises are prevented.

A few years later, in the *Grundrisse*, Marx was writing of "the violent destruction of capital not by relations external to it, but rather as a condition of its self-preservation". In other words, he establishes a necessary link between the generative or creative forces of production in capitalism and the destruction of capital value as one of the key ways in which capitalism attempts to overcome its internal contradictions:

> These contradictions lead to explosions, cataclysms, crises, in which [...] momentaneous suspension of labour and annihilation of a great portion of capital [...] violently lead it back to the point where it is enabled [to go on] fully employing its productive powers without committing suicide.

In the *Theories of Surplus Value* ("Volume IV" of *Das Kapital*, 1863), Marx refines this theory to distinguish between scenarios where the destruction of (commodity) values affects either use values or exchange values or both together. The destruction of exchange value combined with the preservation of use value presents clear opportunities for new capital investment and hence for the repetition of the production-devaluation cycle:

> the destruction of capital through crises means the depreciation of values which prevents them from later renewing their reproduction process as capital on the same scale. This is the ruinous effect of the fall in the prices of commodities. It does not cause the destruction of any use-values. What one loses, the other gains. Values used as capital are prevented from acting again as capital in the hands of the same person. The old capitalists go bankrupt. [...] A large part of the nominal capital of the society, i.e., of the exchange-value of the existing capital, is once for all destroyed, although this very destruction, since it does not affect the use-value, may very much expedite the new reproduction. This is also the period during which moneyed interest enriches itself at the cost of industrial interest.

Social geographer David Harvey sums up the differences between Marx's usage of these concepts and Schumpeter's: "Both Karl Marx and Joseph Schumpeter wrote at length on the 'creative-destructive' tendencies inherent in capitalism. While Marx clearly admired capitalism's creativity he [...] strongly emphasised its self-destructiveness. The Schumpeterians have all along gloried in capitalism's endless creativity while treating the destructiveness as mostly a matter of the normal costs of doing business".

Other Early Usage

In philosophical terms, the concept of "creative destruction" is close to Hegel´s concept of sublation. In German economic discourse it was taken up from Marx's writings by Werner Sombart, particularly in his 1913 text *Krieg und Kapitalismus*:

> Again, however, *from destruction a new spirit of creation arises;* the scarcity of wood and the needs of everyday life... forced the discovery or invention of substitutes for wood, forced the use of coal for heating, forced the invention of coke for the production of iron.

In Hinduism, the god Shiva is simultaneously destroyer and creator, portrayed as *Shiva Nataraja* (Lord of the Dance), which is proposed as the source of the Western notion of "creative destruction".

Hugo Reinert has claimed that Sombart's formulation of the concept was influenced by Eastern mysticism, specifically the image of the Hindu god Shiva, who is presented in the paradoxical aspect of simultaneous destroyer and creator. Conceivably this influence passed from Johann Gottfried Herder, who brought Hindu thought to German philosophy in his *Philosophy of Human History* (Ideen zur Philosophie der Geschichte der Menschheit) (Herder 1790–92), specifically volume III, pp. 41–64. via Arthur Schopenhauer and the Orientalist Friedrich Maier through Friedrich Nietzsche´s writings. Nietzsche represented the creative destruction of modernity through the mythical figure of Dionysus, a figure whom he saw as at one and the same time "destructively creative" and "creatively destructive". In the following passage from *On the Genealogy of Morality* (1887), Nietzsche argues for a universal principle of a cycle of creation and destruction, such that every creative act has its destructive consequence:

> But have you ever asked yourselves sufficiently how much the erection of every ideal on earth has cost? How much reality has had to be misunderstood and slandered, how many lies have had to be sanctified, how many consciences disturbed, how much "God" sacrificed every time? If a temple is to be erected a temple must be destroyed: that is the law – let anyone who can show me a case in which it is not fulfilled! – Friedrich Nietzsche, *On the Genealogy of Morality*

Other nineteenth-century formulations of this idea include Russian anarchist Mikhail Bakunin, who wrote in 1842, "The passion for destruction is a creative passion, too!"

Note, however, that this earlier formulation might more accurately be termed "destructive creation", and differs sharply from Marx's and Schumpeter's formulations in its focus on the active destruction of the existing social and political order by human agents (as opposed to systemic forces or contradictions in the case of both Marx and Schumpeter).

Association with Joseph Schumpeter

The expression "creative destruction" was popularized by and is most associated with Joseph Schumpeter, particularly in his book *Capitalism, Socialism and Democracy*, first published in 1942. Already in his 1939 book *Business Cycles*, he attempted to refine the innovative ideas of Nikolai Kondratieff and his long-wave cycle which Schumpeter believed was driven by technological innovation. Three years later, in *Capitalism, Socialism and Democracy*, Schumpeter introduced the term "creative destruction", which he explicitly derived from Marxist thought (analysed extensively in Part I of the book) and used it to describe the disruptive process of transformation that accompanies such innovation:

> Capitalism [...] is by nature a form or method of economic change and not only never is but never can be stationary. [...] The fundamental impulse that sets and keeps the capitalist engine in motion comes from the new consumers' goods, the new methods of production or transportation, the new markets, the new forms of industrial organization that capitalist enterprise creates.
>
> [...] The opening up of new markets, foreign or domestic, and the organizational development from the craft shop and factory to such concerns as U.S. Steel illustrate the process of industrial mutation that incessantly revolutionizes the economic structure *from within*, incessantly destroying the old one, incessantly creating a new one. This process of Creative Destruction is the essential fact about capitalism. It is what capitalism consists in and what every capitalist concern has got to live in.
>
> [... Capitalism requires] the perennial gale of Creative Destruction.

In Schumpeter's vision of capitalism, innovative entry by entrepreneurs was the disruptive force that sustained economic growth, even as it destroyed the value of established companies and laborers that enjoyed some degree of monopoly power derived from previous technological, organizational, regulatory, and economic paradigms. However, Schumpeter was pessimistic about the sustainability of this process, seeing it as leading eventually to the undermining of capitalism's own institutional frameworks:

> In breaking down the pre-capitalist framework of society, capitalism thus broke not only barriers that impeded its progress but also flying buttresses that prevented its collapse. That process, impressive in its relentless necessity, was not merely a matter of removing institutional deadwood, but of removing partners

> of the capitalist stratum, symbiosis with whom was an essential element of the capitalist schema. [... T]he capitalist process in much the same way in which it destroyed the institutional framework of feudal society also undermines its own.

Schumpeter nevertheless elaborated the concept, making it central to his economic theory, and it was later taken up as a major doctrine of the so-called Austrian School of free-market economic thought.

Examples

Polaroid instant cameras have disappeared almost completely with the spread of digital photography.

Schumpeter (1949) in one of his examples used "the railroadization of the Middle West as it was initiated by the Illinois Central." He wrote, "The Illinois Central not only meant very good business whilst it was built and whilst new cities were built around it and land was cultivated, but it spelled the death sentence for the [old] agriculture of the West."

Companies that once revolutionized and dominated new industries – for example, Xerox in copiers or Polaroid in instant photography – have seen their profits fall and their dominance vanish as rivals launched improved designs or cut manufacturing costs. In technology, the cassette tape replaced the 8-track, only to be replaced in turn by the compact disc, which was undercut by downloads to MP3 players, which is now being usurped by web-based streaming services. Companies which made money out of technology which becomes obsolete do not necessarily adapt well to the business environment created by the new technologies.

One such example is the way in which online ad-supported news sites such as *The Huffington Post* are leading to creative destruction of the traditional newspaper. The *Christian Science Monitor* announced in January 2009 that it would no longer continue to publish a daily paper edition, but would be available online daily and provide a weekly print edition. The *Seattle Post-Intelligencer* became online-only in March

2009. At a national level, employment in the newspaper business fell from 455,700 in 1990 to 225,100 in 2013. Over that same period, employment in internet publishing and broadcasting grew from 29,400 to 121,200. Traditional French alumni networks, which typically charge their students to network online or through paper directories, are in danger of creative destruction from free social networking sites such as Linkedin and Viadeo.

In fact, successful innovation is normally a source of temporary market power, eroding the profits and position of old firms, yet ultimately succumbing to the pressure of new inventions commercialised by competing entrants. Creative destruction is a powerful economic concept because it can explain many of the dynamics or kinetics of industrial change: the transition from a competitive to a monopolistic market, and back again. It has been the inspiration of endogenous growth theory and also of evolutionary economics.

David Ames Wells (1890), who was a leading authority on the effects of technology on the economy in the late 19th century, gave many examples of creative destruction (without using the term) brought about by improvements in steam engine efficiency, shipping, the international telegraph network, and agricultural mechanization.

Later Developments

Ludwig Lachmann

These economic facts have certain social consequences. As the critics of the market economy nowadays prefer to take their stand on "social" grounds, it may be not inappropriate here to elucidate the true social results of the market process. We have already spoken of it as a leveling process. More aptly, we may now describe these results as an instance of what Pareto called "the circulation of elites." Wealth is unlikely to stay for long in the same hands. It passes from hand to hand as unforeseen change confers value, now on this, now on that specific resource, engendering capital gains and losses. The owners of wealth, we might say with Schumpeter, are like the guests at a hotel or the passengers in a train: They are always there but are never for long the same people.

—Ludwig Lachmann, The Market Economy and the Distribution of Wealth

David Harvey

Geographer and historian David Harvey in a series of works from the 1970s onwards (*Social Justice and the City*, 1973; *The Limits to Capital*, 1982; *The Urbanization of Capital*, 1985; *Spaces of Hope*, 2000; *Spaces of Capital*, 2001; *Spaces of Neoliberalization*, 2005; *The Enigma of Capital and the Crises of Capitalism*, 2010), elaborated Marx's thought on the systemic contradictions of capitalism, particularly in relation to the production of the urban environment (and to the production of space more broadly). He developed the notion that capitalism finds a "spatial fix" for its periodic crises of overaccumulation through investment in fixed assets of infrastructure, buildings, etc.: "The built environment that constitutes a vast field of collective means of production and consumption absorbs huge amounts of capital in both its construction and its maintenance. Urbanization is one way to absorb the capital surplus". While the creation of the built environment can act as a form of crisis displacement, it can also constitute a limit in its own right, as it tends to freeze productive forces into a fixed spatial form. As capital cannot abide a limit to profitability, ever more frantic forms of "time-space compression" (increased speed of turnover, innovation of ever faster transport and communications' infrastructure, "flexible accumulation") ensue, often impelling technological innovation. Such innovation, however, is a double-edged sword:

The effect of continuous innovation [...] is to devalue, if not destroy, past investments and labour skills. *Creative destruction* is embedded within the circulation of capital itself. Innovation exacerbates instability, insecurity, and in the end, becomes the prime force pushing capitalism into periodic paroxysms of crisis. [...] The struggle to maintain profitability sends capitalists racing off to explore all kinds of other possibilities. New product lines are opened up, and that means the creation of new wants and needs. Capitalists are forced to redouble their efforts to create new needs in others [...]. The result is to exacerbate insecurity and instability, as masses of capital and workers shift from one line of production to another, leaving whole sectors devastated [...]. The drive to relocate to more advantageous places (the geographical movement of both capital and labour) periodically revolutionizes the international and territorial division of labour, adding a vital geographical dimension to the insecurity. The resultant transformation in the experience of space and place is matched by revolutions in the time dimension, as capitalists strive to reduce the turnover time of their capital to "the twinkling of an eye".

Globalization can be viewed as some ultimate form of time-space compression, allowing capital investment to move almost instantaneously from one corner of the globe to another, devaluing fixed assets and laying off labour in one urban conglommeration while opening up new centres of manufacture in more profitable sites for production operations. Hence, in this continual process of creative destruction, capitalism does not resolve its contradictions and crises, but merely "moves them around geographically".

Marshall Berman

In his 1987 book *All That is Solid Melts into Air: The Experience of Modernity*, particularly in the chapter entitled "Innovative Self-Destruction" (pp. 98–104), Marshall Berman provides a reading of Marxist "creative destruction" to explain key processes at work within modernity. The title of the book is taken from a well-known passage from *The Communist Manifesto*. Berman elaborates this into something of a *Zeitgeist* which has profound social and cultural consequences:

The truth of the matter, as Marx sees, is that everything that bourgeois society builds is built to be torn down. "All that is solid" — from the clothes on our backs to the looms and mills that weave them, to the men and women who work the machines, to the houses and neighborhoods the workers live in, to the firms and corporations that exploit the workers, to the towns and cities and whole regions and even nations that embrace them all — all these are made to be broken tomorrow, smashed or shredded or pulverized or dissolved, so they can be recycled or replaced next week, and the whole process can go on again and again, hopefully forever, in ever more profitable forms. The pathos of all bourgeois monuments is that their material strength and solidity actually count for nothing and carry no weight at all, that they are blown away like frail reeds by the very forces of capitalist development that they celebrate. Even the most beautiful and impressive bourgeois buildings and public works are disposable, capitalized for fast depreciation and planned to be obsolete, closer in their social functions to tents and encampments than to "Egyptian pyramids, Roman aqueducts, Gothic cathedrals".

Here Berman emphasizes Marx's perception of the fragility and evanescence of capitalism's immense creative forces, and makes this apparent contradiction into one of the key explanatory figures of modernity.

Manuel Castells

The sociologist Manuel Castells, in his trilogy on *The Information Age: Economy, Society and Culture* (the first volume of which, *The Rise of the Network Society*, appeared in 1996), reinterpreted the processes by which capitalism invests in certain regions of the globe, while divesting from others, using the new paradigm of "informational networks". In the era of globalization, capitalism is characterized by near-instantaneous flow, creating a new spatial dimension, "the space of flows". While technological innovation has enabled this unprecedented fluidity, this very process makes redundant whole areas and populations who are bypassed by informational networks. Indeed, the new spatial form of the mega-city or megalopolis, is defined by Castells as having the contradictory quality of being "globally connected and locally disconnected, physically and socially". Castells explicitly links these arguments to the notion of creative destruction:

The "spirit of informationalism" is the culture of "creative destruction" accelerated to the speed of the optoelectronic circuits that process its signals. Schumpeter meets Weber in the cyberspace of the network enterprise.

Daniele Archibugi

Developing the Schumpeterian legacy, the school of the Science Policy Research Unit of the University of Sussex has further detailed the importance of creative destruction exploring, in particular, how new technologies are often idiosyncratic with the existing productive regimes and will lead to bankruptcy companies and even industries that do not manage to sustain the rate of change. Chris Freeman and Carlota Perez have developed these insights. More recently, Daniele Archibugi and Andrea Filippetti have associated the 2008 economic crisis to the slow-down of opportunities offered by information and communication technologies (ICTs). Using as a metaphor the film Blade Runner, Archibugi has argued that of the innovations described in the film in 1982, all those associated to ICTs have become part of our everyday life. But, on the contrary, none of those in the field of Biotech have been fully commercialized. A new economic recovery will occur when some key technological opportunities will be identified and sustained.

Technological opportunities do not enter into economic and social life without deliberate efforts and choices. We should be able to envisage new forms of organization associated with emerging technology. ICTs have already changed our lifestyle even more than our economic life: they have generated jobs and profits, but above all they have transformed the way we use our time and interact with the world. Biotech could bring about even more radical social transformations at the core of our life. Why have these not yet been delivered? What can be done to unleash their potential? There are a few basic questions that need to be addressed.

Others

In 1992, the idea of creative destruction was put into formal mathematical terms by Philippe Aghion and Peter Howitt, giving an alternative model of endogenous growth compared to Paul Romer's expanding varieties model.

In 1995, Harvard Business School authors Richard L. Nolan and David C. Croson released *Creative Destruction: A Six-Stage Process for Transforming the Organization.* The book advocated downsizing to free up slack resources, which could then be reinvested to create competitive advantage.

More recently, the idea of "creative destruction" was utilized by Max Page in his 1999 book, *The Creative Destruction of Manhattan, 1900–1940.* The book traces Manhattan's constant reinvention, often at the expense of preserving a concrete past. Describing this process as "creative destruction," Page describes the complex historical circumstances, economics, social conditions and personalities that have produced crucial changes in Manhattan's cityscape.

In addition to Max Page, others have used the term "creative destruction" to describe the process of urban renewal and modernization. T.C. Chang and Shirlena Huang referenced "creative destruction" in their paper *Recreating place, replacing memory: Creative Destruction at the Singapore River.* The authors explored the efforts to redevelop a waterfront area that reflected a vibrant new culture while paying sufficient homage to the history of the region. Rosemary Wakeman chronicled the evolution of an area in central Paris, France known as Les Halles. Les Halles housed a vibrant marketplace starting in the twelfth century. Ultimately, in 1971, the markets were relocated and the pavilions torn down. In their place, now stand a hub for trains, subways and buses. Les Halles is also the site of the largest shopping mall in France and the controversial Centre Georges Pompidou.

The term "creative destruction" has been applied to the arts. Alan Ackerman and Martin Puncher (2006) edited a collection of essays under the title *Against Theater: Creative destruction on the modernist stage*. They detail the changes and the causal motivations experienced in theater as a result of the modernization of both the production of performances and the underlying economics. They speak of how theater has reinvented itself in the face of anti-theatricality, straining the boundaries of the traditional to include more physical productions, which might be considered avant-garde staging techniques.

Neoconservative author Michael Ledeen argued in his 2002 book *The War Against the Terror Masters* that America is a revolutionary nation, undoing traditional societies: "Creative destruction is our middle name, both within our own society and abroad. We tear down the old order every day, from business to science, literature, art, architecture, and cinema to politics and the law." His characterization of creative destruction as a model for social development has met with fierce opposition from paleoconservatives.

Creative destruction has also been linked to sustainable development. The connection was explicitly mentioned for the first time by Stuart L. Hart and Mark B. Milstein in their 1999 article *Global Sustainability and the Creative Destruction of Industries*, in which he argues new profit opportunities lie in a round of creative destruction driven by global sustainability. (An argument which they would later on strengthen in their 2003 article *Creating Sustainable Value* and, in 2005, with *Innovation, Creative Destruction and Sustainability*.) Andrea L. Larson agreed with this vision a year later in *Sustainable Innovation Through an Entrepreneurship Lens*, stating entrepreneurs should be open to the opportunities for disruptive improvement based on sustainability. In 2005, James Hartshorn (et al.) emphasized the opportunities for sustainable, disruptive improvement in the construction industry in his article *Creative Destruction: Building Toward Sustainability*.

Alternative Name

The following text appears to be the source of the phrase "Schumpeter's Gale" to refer to creative destruction:

The opening up of new markets and the organizational development from the craft shop and factory to such concerns as US Steel illustrate the process of industrial mutation that incessantly revolutionizes the economic structure from within, incessantly destroying the old one, incessantly creating a new one ... [The process] must be seen in its role in the perennial gale of creative destruction; it cannot be understood on the hypothesis that there is a perennial lull.

—Joseph Schumpeter, Capitalism, Socialism and Democracy, 1942

In Popular Culture

The film *Other People's Money* (1991) provides contrasting views of creative destruction, presented in two speeches regarding the takeover of a publicly traded wire and cable company in a small New England town. One speech is by a corporate raider, and the other is given by the company CEO, who is principally interested in protecting his employees and the town.

Pro-innovation Bias

In diffusion of innovation theory, a pro-innovation bias is the belief that an innovation should be adopted by whole society without the need of its alteration. The innovation's "champion" has such strong bias in favor of the innovation, that he/she may not see its limitations or weaknesses and continues to promote it nonetheless.

Example

A feeling of nuclear optimism emerged in the 1950s in which it was believed that all power generators in the future would be atomic in nature. The atomic bomb would render all conventional explosives obsolete and nuclear power plants would do the same for power sources such as coal and oil. There was a general feeling that everything would use a nuclear power source of some sort, in a positive and productive way, from irradiating food to preserve it, to the development of nuclear medicine. There would be an age of peace and plenty in which atomic energy would "provide the power needed to desalinate water for the thirsty, irrigate the deserts for the hungry, and fuel interstellar travel deep into outer space". This use would render the Atomic Age as significant a step in technological progress as the first smelting of Bronze, of Iron, or the commencement of the Industrial Revolution.

Roger Smith, then chairman of General Motors, said in 1986: "By the turn of the century, we will live in a paperless society." In the late 20th century, there were many predictions of this kind. This transformation has so far not taken place.

References

- Sutherland, Stuart (2007), Irrationality (2nd ed.), London: Pinter and Martin, pp. 95–103, ISBN 978-1-905177-07-3, OCLC 72151566
- Barkow, Jerome H.; Cosmides, Leda; Tooby, John (1995), The adapted mind: evolutionary psychology and the generation of culture, Oxford University Press US, pp. 181–184, ISBN 978-0-19-510107-2, OCLC 33832963
- Matlin, Margaret W. (2004), "Pollyanna Principle", in Pohl, Rüdiger F., Cognitive Illusions: A Handbook on Fallacies and Biases in Thinking, Judgement and Memory, Hove: Psychology Press, pp. 255–272, ISBN 978-1-84169-351-4, OCLC 55124398

- Shanteau, James (2003). Sandra L. Schneider, ed. Emerging perspectives on judgment and decision research. Cambridge [u.a.]: Cambridge University Press. p. 445. ISBN 0-521-52718-X.
- Haidt, Jonathan (2012). The Righteous Mind : Why Good People are Divided by Politics and Religion. New York: Pantheon Books. pp. 1473–4 (e–book edition). ISBN 0-307-37790-3.
- Lindzey, edited by Susan T. Fiske, Daniel T. Gilbert, Gardner (2010). The handbook of social psychology. (5th ed.). Hoboken, N.J.: Wiley. p. 811. ISBN 0-470-13749-5.
- Pompian, Michael M. (2006), Behavioral finance and wealth management: how to build optimal portfolios that account for investor biases, John Wiley and Sons, pp. 187–190, ISBN 978-0-471-74517-4, OCLC 61864118
- Singh, Simon; Ernst, Edzard (2008), Trick or Treatment?: Alternative Medicine on Trial, London: Bantam, pp. 287–288, ISBN 978-0-593-06129-9
- Neenan, Michael; Dryden, Windy (2004), Cognitive therapy: 100 key points and techniques, Psychology Press, p. ix, ISBN 978-1-58391-858-6, OCLC 474568621
- Blackburn, Ivy-Marie; Davidson, Kate M. (1995), Cognitive therapy for depression & anxiety: a practitioner's guide (2 ed.), Wiley-Blackwell, p. 19, ISBN 978-0-632-03986-9, OCLC 32699443
- Halpern, Diane F. (1987), Critical thinking across the curriculum: a brief edition of thought and knowledge, Lawrence Erlbaum Associates, p. 194, ISBN 978-0-8058-2731-6, OCLC 37180929
- Tetlock, Philip E. (2005), Expert Political Judgment: How Good Is It? How Can We Know?, Princeton, N.J.: Princeton University Press, pp. 125–128, ISBN 978-0-691-12302-8, OCLC 56825108
- Proctor, Robert W.; Capaldi, E. John (2006), Why science matters: understanding the methods of psychological research, Wiley-Blackwell, p. 68, ISBN 978-1-4051-3049-3, OCLC 318365881
- Christensen, Clayton M. (2003). The innovator's solution : creating and sustaining successful growth. Harvard Business Press. ISBN 978-1-57851-852-4.

4

Theories and Models of Innovation Management

The process of understanding the response of the population towards new products is known as the bass diffusion model. Along with the bass model, the theories and models of innovation management explained in this chapter are actor-network theory, diffusion of innovations, induced innovation, social theory, chain-linked model etc.

Actor–network Theory

Actor–network theory (ANT) is an approach to social theory and research, originating in the field of science studies, which treats objects as part of social networks. Although it is best known for its controversial insistence on the capacity of nonhumans to act or participate in systems or networks or both, ANT is also associated with forceful critiques of conventional and critical sociology. Developed by science and technology studies (STS) scholars Michel Callon and Bruno Latour, the sociologist John Law, and others, it can more technically be described as a "material-semiotic" method. This means that it maps relations that are simultaneously material (between things) and semiotic (between concepts). It assumes that many relations are both material and semiotic.

Broadly speaking, ANT is a constructivist approach in that it avoids essentialist explanations of events or innovations (i. e. ANT explains a successful theory by understanding the combinations and interactions of elements that make it successful, rather than saying it is "true" and the others are "false"). However, it is distinguished from many other STS and sociological network theories for its distinct material-semiotic approach.

Background and Context

ANT was first developed at the Centre de Sociologie de l'Innovation (CSI) of the École nationale supérieure des mines de Paris in the early 1980s by staff (Michel Callon and Bruno Latour) and visitors (including John Law). The 1984 book co-authored by John Law and fellow-sociologist Peter Lodge (*Science for Social Scientists*; London: Macmillan Press Ltd.) is a good example of early explorations of how the growth and structure of knowledge could be analyzed and interpreted through the interactions of actors and networks. Initially created in an attempt to understand processes of innovation and knowledge-creation in science and technology, the approach drew on existing work in

STS, on studies of large technological systems, and on a range of French intellectual resources including the semiotics of Algirdas Julien Greimas, the writing of philosopher Michel Serres, and the Annales School of history.

ANT appears to reflect many of the preoccupations of French post-structuralism, and in particular a concern with non-foundational and multiple material-semiotic relations. At the same time, it was much more firmly embedded in English-language academic traditions than most post-structuralist-influenced approaches. Its grounding in (predominantly English) science and technology studies was reflected in an intense commitment to the development of theory through qualitative empirical case-studies. Its links with largely US-originated work on large technical systems were reflected in its willingness to analyse large scale technological developments in an even-handed manner to include political, organizational, legal, technical and scientific factors.

Many of the characteristic ANT tools (including the notions of translation, generalized symmetry and the "heterogeneous network"), together with a scientometric tool for mapping innovations in science and technology ("co-word analysis") were initially developed during the 1980s, predominantly in and around the CSI. The "state of the art" of ANT in the late 1980s is well-described in Latour's 1987 text, *Science in Action*.

From about 1990 onwards, ANT started to become popular as a tool for analysis in a range of fields beyond STS. It was picked up and developed by authors in parts of organizational analysis, informatics, health studies, geography, sociology, anthropology, feminist studies and economics.

As of 2008, ANT is a widespread, if controversial, range of material-semiotic approaches for the analysis of heterogeneous relations. In part because of its popularity, it is interpreted and used in a wide range of alternative and sometimes incompatible ways. There is no orthodoxy in current ANT, and different authors use the approach in substantially different ways. Some authors talk of "after-ANT" to refer to "successor projects" blending together different problem-focuses with those of ANT.

A material-semiotic Method

Although it is called a "theory", ANT does not usually explain "why" or "how" a network takes the form that it does. Rather, ANT is a way of thoroughly exploring the relational ties within a network (which can be a multitude of different things). As Latour notes, "explanation does not follow from description; it is description taken that much further." It is not, in other words, a theory "of" anything, but rather a method, or a "how-to book" as Latour puts it.

The approach is related to other versions of material-semiotics (notably the work of philosophers Gilles Deleuze, Michel Foucault and feminist scholar Donna Haraway). It can also be seen as a way of being faithful to the insights of ethnomethodology and its detailed descriptions of how common activities, habits and procedures sustain themselves. Similarities between ANT and symbolic interactionist approaches such as the newer forms of grounded theory like situational analysis, exist, although Latour objects to such a comparison.

Although ANT is mostly associated with studies of science and technology and with the sociology of science, it has been making steady progress in other fields of sociology as well. ANT is adamantly empirical, and as such yields useful insights and tools for sociological inquiry in general. ANT has been deployed in studies of identity and subjectivity, urban transportation systems, and passion and addiction. It also makes steady progress in political and historical sociology.

The Actor-network

As the term implies, the actor-network is the central concept in ANT. The term "network" is somewhat problematic in that it, as Latour notes, has a number of unwanted connotations. Firstly, it implies that what is described takes the shape of a network, which is not necessarily the case. Secondly, it implies "transportation without deformation," which, in ANT, is not possible since any actor-network involves a vast number of translations. Latour, however, still contends that network is a fitting term to use, because "it has no a priori order relation; it is not tied to the axiological myth of a top and of a bottom of society; it makes absolutely no assumption whether a specific locus is macro- or micro- and does not modify the tools to study the element 'a' or the element 'b'." This use of the term "network" is very similar to Deleuze and Guattari's rhizomes; Latour even remarks tongue in cheek that he would have no objection to renaming ANT "actant-rhizome ontology" if it only had sounded better, which hints at Latour's uneasiness with the word "theory".

Actor–network theory tries to explain how material–semiotic networks come together to act as a whole; the clusters of actors involved in creating meaning are both material and semiotic. As a part of this it may look at explicit strategies for relating different elements together into a network so that they form an apparently coherent whole. These networks are potentially transient, existing in a constant making and re-making. This means that relations need to be repeatedly "performed" or the network will dissolve. They also assume that networks of relations are not intrinsically coherent, and may indeed contain conflicts. Social relations, in other words, are only ever in process, and must be performed continuously.

Actants denote human and non-human actors, and in a network take the shape that they do by virtue of their relations with one another. It assumes that nothing lies outside the network of relations, and as noted above, suggests that there is no difference in the ability of technology, humans, animals, or other non-humans to act (and that there are only enacted alliances). As soon as an actor engages with an actor-network it too is caught up in the web of relations, and becomes part of the entelechy.

If taken to its logical conclusion, then, nearly any actor can be considered merely a sum of other, smaller actors. A car is an example of a complex system. It contains many electronic and mechanical components, all of which are essentially hidden from view to the driver, who simply deals with the car as a single object. This effect is known as *punctualisation*, and is similar to the idea of encapsulation in object-oriented programming.

When an actor network breaks down, the punctualisation effect tends to cease as well. In the automobile example above, a non-working engine would cause the driver to become aware of the car as a collection of parts rather than just a vehicle capable of transporting him or her from place to place. This can also occur when elements of a network act contrarily to the network as a whole. In his book *Pandora's Hope*, Latour likens depunctualization to the opening of a black box. When closed, the box is perceived simply as a box, although when it is opened all elements inside it becomes visible.

Human and non-human Actors

ANT is often associated with the equal treatment of human and non-human actors. ANT assumes that all entities in a network can and should be described in the same terms. This is called the principle of *generalized symmetry*. The rationale for this is that differences between them are generated in the network of relations, and should not be presupposed.

Intermediaries and Mediators

The distinction between intermediaries and mediators is key to ANT sociology. Intermediaries are entities which make no difference (to some interesting state of affairs which we are studying) and so can be ignored. They transport the force of some other entity more or less without transformation and so are fairly uninteresting. Mediators are entities which multiply difference and so should be the object of study. Their outputs cannot be predicted by their inputs. From an ANT point of view sociology has tended to treat too much of the world as intermediaries.

For instance, a sociologist might take silk and nylon as intermediaries, holding that the former "means", "reflects", or "symbolises" the upper classes and the latter the lower classes. In such a view the real world silk–nylon difference is irrelevant — presumably many other material differences could also, and do also, transport this class distinction. But taken as mediators these fabrics would have to be engaged with by the analyst in their specificity: the internal real-world complexities of silk and nylon suddenly appear relevant, and are seen as actively constructing the ideological class distinction which they once merely reflected.

For the committed ANT analyst, social things—like class distinctions in taste in the silk and nylon example, but also groups and power—must constantly be constructed or performed anew through complex engagements with complex mediators. There is no stand-alone social repertoire lying in the background to be reflected off, expressed through, or substantiated in, interactions (as in an intermediary conception).

Other Central Concepts

Translation

Central to ANT is the concept of translation which is sometimes referred to as sociology of translation, in which innovators attempt to create a *forum,* a central network in which all the actors agree that the network is worth building and defending. In his widely debated 1986 study of how marine biologists try to restock the St Brieuc Bay in order to produce more scallops, Michel Callon has defined 4 moments of translation: problematization, interessement, enrollment and mobilisation of allies. Also important to the notion is the role of network objects in helping to smooth out the translation process by creating equivalencies between what would otherwise be very challenging people, organizations or conditions to mesh together. Bruno Latour spoke about this particular task of objects in his work *Reassembling the Social* (2005).

Tokens, or Quasi-objects

In the above examples, "social order" and "functioning car" come into being through the successful interactions of their respective actor-networks, and actor-network theory refers to these creations as *tokens* or *quasi-objects* which are passed between actors within the network.

As the token is increasingly transmitted or passed through the network, it becomes increasingly punctualized and also increasingly reified. When the token is decreasingly transmitted, or when an actor fails to transmit the token (e.g., the oil pump breaks), punctualization and reification are decreased as well.

Actor Network Theory and Specific Disciplines

Recently, there has been a movement to introduce Actor Network Theory as an analytical tool to a range of applied disciplines outside of sociology, including nursing, public health, business (Klein and Czarniawska, 2005), urban studies (Farias and Bender, 2010), and community, urban, and regional planning (Beauregard, 2012; Beauregard and Lieto, 2015; Rydin, 2012; Rydin and Tate, 2016, Tate, 2013.).

ANT and International Relations

Actor-Network Theory has become increasingly prominent within the discipline of International Relations and Political Science.

Theoretically, scholars within IR have employed ANT in order to disrupt traditional world political binaries (civilised/barbarian, democratic/autocratic, etc.), consider the implications of a posthuman understanding of IR, explore the infrastructures of world politics, and consider the effects of technological agency.

Empirically, IR scholars have drawn on insights from ANT in order to study phenomena including political violences like the use of torture and drones, piracy and maritime governance, and garbage.

ANT and Design

The Actor-Network Theory can also be applied to design, using a perspective that is not simply limited to an analysis of an object's structure. From the ANT viewpoint, design is seen as a series of features that account for a social, psychological, and economical world. ANT argues that objects are designed to shape human action and mold or influence decisions. In this way, the objects' design serves to mediate human relationships and can even impact our morality, ethics, and politics.

Criticism

Actor–network theory insists on the capacity of nonhumans to be actors or participants in networks and systems. Critics including figures such as Langdon Winner maintain that such properties as *intentionality* fundamentally distinguish humans from animals or from "things". ANT scholars respond with the following arguments:

- They do not attribute intentionality and similar properties to nonhumans.
- Their conception of agency does not presuppose intentionality.
- They locate agency neither in human "subjects" nor in non-human "objects", but in heterogeneous associations of humans and nonhumans.

ANT has been criticized as amoral. Wiebe Bijker has responded to this criticism by stating that the amorality of ANT is not a necessity. Moral and political positions are possible, but one must first describe the network before taking up such positions. This position has been further explored by Stuart Shapiro who contrasts ANT with the history of ecology, and argues that research decisions are moral rather than methodological, but this moral dimension has been sidelined.

Whittle and Spicer note that "ANT has also sought to move beyond deterministic models that trace organizational phenomena back to powerful individuals, social structures, hegemonic discourses or technological effects. Rather, ANT prefers to seek out complex patterns of causality rooted in connections between actors." They argue that ANT's ontological realism makes it "less well equipped for pursuing a critical account of organizations—that is, one which recognises the unfolding nature of reality, considers the limits of knowledge and seeks to challenge structures of domination." This implies that ANT does not account for pre-existing structures, such as power, but rather sees these structures as emerging from the actions of actors within the network and their ability to align in pursuit of their interests. Accordingly, ANT can be seen as an attempt to re-introduce Whig history into science and technology studies; like the myth of the heroic inventor, ANT can be seen as an attempt to explain successful innovators by saying only that they were successful. Likewise, for organization studies, Whittle and Spicer assert that ANT is, "ill suited to the task of developing political alternatives to the imaginaries of market managerialism."

Key early criticism came from other members of the STS community, in particular the "Epistemological Chicken" debate between Collins and Yearley with responses from Latour and Callon as well as Woolgar. Collins and Yearley accused ANTs approach of collapsing into an endless relativist regress. Some critics have argued that research based on ANT perspectives remains entirely descriptive and fails to provide explanations for social processes. ANT—like comparable social scientific methods—requires judgement calls from the researcher as to which actors are important within a network and which are not. Critics argue that the importance of particular actors cannot be determined in the absence of "out-of-network" criteria. Similarly, others argue that actor-networks risk degenerating into endless chains of association (six degrees of separation—we are all networked to one another). Other research perspectives such as social constructionism, social shaping of technology, social network theory, normalization process theory, Diffusion of Innovations theory are held to be important alternatives to ANT approaches.

In a workshop called "On Recalling ANT", Bruno Latour stated that there are four things wrong with actor-network theory: "actor", "network", "theory" and the hyphen. In a later book however (*Reassembling the Social: An Introduction to Actor–Network–Theory*), Latour reversed himself, accepting the wide use of the term, "*including* the hyphen" (Latour 2005:9). He also remarked how he had been helpfully reminded that the ANT acronym "was perfectly fit for a blind, myopic, workaholic, trail-sniffing, and collective traveler" (the ant, Latour 2005:9) — qualitative hallmarks of actor-network epistemology.

Bass Diffusion Model

The Bass Model or Bass Diffusion Model was developed by Frank Bass. It consists of a simple differential equation that describes the process of how new products get adopted in a population. The model presents a rationale of how current adopters and potential adopters of a new product interact. The basic premise of the model is that adopters can be classified as innovators or as imitators and the speed and timing of adoption depends on their degree of innovativeness and the degree of imitation among adopters. The Bass model has been widely used in forecasting, especially new products' sales forecasting and technology forecasting. Mathematically, the basic Bass diffusion is a Riccati equation with constant coefficients.

Frank Bass published his paper "A new product growth for model consumer durables" in 1969 whose title indeed contained a typographical error. Prior to this, Everett Rogers published *Diffusion of Innovations*, a highly influential work that described the different stages of product adoption. Bass contributed some mathematical ideas to the concept.

Model Formulation

$$\frac{f(t)}{1-F(t)} = p + qF(t)$$

Where:

- $f(t)$ is the change of the installed base fraction
- $F(t)$ is the installed base fraction
- is the coefficient of innovation
- q is the coefficient of imitation

Sales $S(t)$ is the rate of change of installed base (i.e. adoption) $f(t)$ multiplied by the ultimate market potential m:

$$S(t) = mf(t)$$

$$S(t) = m\frac{(p+q)^2}{p}\frac{e^{-(p+q)t}}{(1+\frac{q}{p}e^{-(p+q)t})^2}$$

The time of peak sales t^*

$$t^* = \frac{\ln q - \ln p}{p+q}$$

Explanation

The coefficient p is called the coefficient of innovation, external influence or advertising effect. The coefficient q is called the coefficient of imitation, internal influence or word-of-mouth effect.

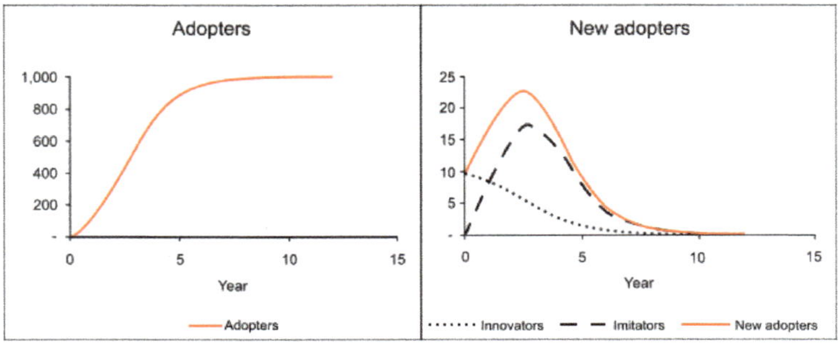

Typical values of p and q when time t is measured in years:

- The average value of p has been found to be 0.03, and is often less than 0.01

- The average value of q has been found to be 0.38, with a typical range between 0.3 and 0.5

Extensions to the model

Generalised Bass model (with pricing)

Bass found that his model fit the data for almost all product introductions, despite a wide range of managerial decision variables, e.g. pricing and advertising. This means that decision variables can shift the Bass curve in time, but that the shape of the curve is always similar.

Although many extensions of the model have been proposed, only one of these reduces to the Bass model under ordinary circumstances.

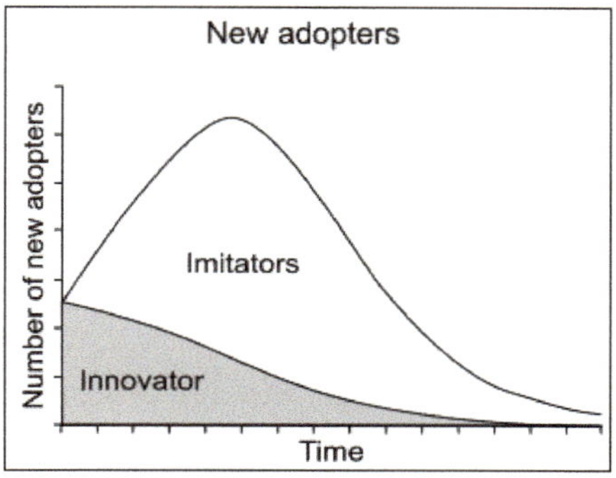

This model was developed in 1994 by Frank Bass, Trichy Krishnan and Dipak Jain:

$$\frac{f(t)}{1-F(t)} = (p + qF(t))x(t)$$

where $x(t)$ is a function of percentage change in price and other variables

Successive Generations

Technology products succeed one another in generations. Norton and Bass extended the model in 1987 for sales of products with continuous repeat purchasing. The formulation for three generations is as follows:

$$S_{1,t} = F(t_1)m_1(1-F(t_2))$$

$$S_{2,t} = F(t_2)(m_2 + F(t_1)m_1)(1-F(t_3))$$

$$S_{3,t} = F(t_3)(m_3 + F(t_2)(m_2 + F(t_1)m_1))$$

where

- $m_i = a_i M_i$
- M_i is the incremental number of ultimate adopters of the ith generation product

- a_i is the average (continuous) repeat buying rate among adopters of the *i*th generation product
- t_i is the time since the introduction of the *i*th generation product
- $$F(t_i) = \frac{1 - e^{-(p+q)t_i}}{1 + \frac{q}{p} e^{-(p+q)t_i}}$$

It has been found that the p and q terms are generally the same between successive generations.

Relationship with other S-curves

There are two special cases of the Bass diffusion model.

- The first special case occurs when q=0, when the model reduces to the Exponential distribution.
- The second special case reduces to the logistic distribution, when p=0.

The Bass model is a special case of the Gamma/shifted Gompertz distribution (G/SG): Bemmaor (1994)

Use in online social networks The rapid, recent (as of early 2007) growth in online social networks (and other virtual communities) has led to an increased use of the Bass diffusion model. The Bass diffusion model is used to estimate the size and growth rate of these social networks. The work by Christian Bauchkage and co-authors shows that the Bass model provides a more optimistic picture of the future than alternative model(s) such as the Weibull distribution and the shifted Gompertz distribution.

Adoption of this Model

The model is one of the most cited empirical generalizations in marketing; as of December 2014 the paper "A New Product Growth for Model Consumer Durables" published in Management Science had (approximately) 5740 citations in Google Scholar.

This model has been widely influential in marketing and management science. In 2004 it was selected as one of the ten most frequently cited papers in the 50-year history of *Management Science*. It was ranked number five, and the only marketing paper in the list. It was subsequently reprinted in the December 2004 issue of Management Science.

Diffusion of Innovations

Diffusion of innovations is a theory that seeks to explain how, why, and at what rate new ideas and technology spread. Everett Rogers, a professor of communication studies, popularized the theory in his book *Diffusion of Innovations*; the book was first published in 1962, and is now in its fifth edition (2003). Rogers argues that diffusion is the process by which an innovation is communicated over time among the participants in a social system. The origins of the diffusion of innovations theory are varied and span multiple disciplines.

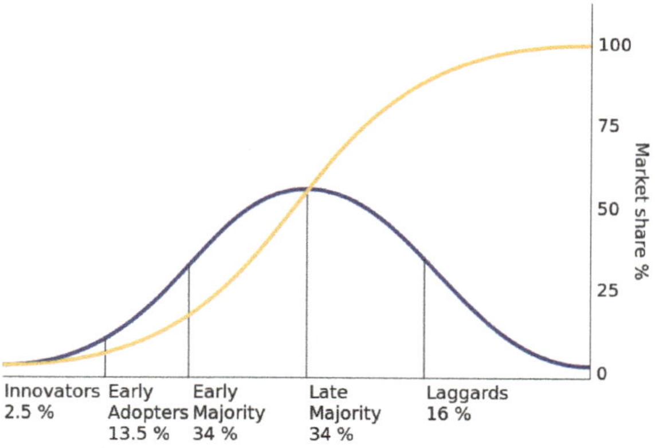

The diffusion of innovations according to Rogers. With successive groups of consumers adopting the new technology (shown in blue), its market share (yellow) will eventually reach the saturation level. In mathematics, the yellow curve is known as the logistic function. The curve is broken into sections of adopters.

Rogers proposes that four main elements influence the spread of a new idea: the innovation itself, communication channels, time, and a social system. This process relies heavily on human capital. The innovation must be widely adopted in order to self-sustain. Within the rate of adoption, there is a point at which an innovation reaches critical mass.

The categories of adopters are innovators, early adopters, early majority, late majority, and laggards. Diffusion manifests itself in different ways and is highly subject to the type of adopters and innovation-decision process. The criterion for the adopter categorization is innovativeness, defined as the degree to which an individual adopts a new idea.

History

The concept of diffusion was first studied by the French sociologist Gabriel Tarde in late 19th century and by German and Austrian anthropologists and geographers such as Friedrich Ratzel and Leo Frobenius. The study of diffusion of innovations took off in the subfield of rural sociology in the midwestern United States in the 1920s and 1930s. Agriculture technology was advancing rapidly, and researchers started to examine how independent farmers were adopting hybrid seeds, equipment, and techniques. A study of the adoption of hybrid corn seed in Iowa by Ryan and Gross (1943) solidified the prior work on diffusion into a distinct paradigm that would be cited consistently in the future. Since its start in rural sociology, Diffusion of Innovations has been applied to numerous contexts, including medical sociology, communications, marketing, development studies, health promotion, organizational studies, knowledge management, and complexity studies, with a particularly large impact on the use of medicines, medical techniques, and health communications. In organizational studies, its basic epidemiological or internal-influence form was formulated by H. Earl Pemberton, who provided examples of institutional diffusion such as postage stamps and standardized school ethics codes.

In 1962, Everett Rogers, a professor of rural sociology, published his seminal work: *Diffusion of Innovations*. Rogers synthesized research from over 508 diffusion studies across the fields that initially influenced the theory: anthropology, early sociology, rural sociology, education, industrial sociology and medical sociology. Using his synthesis, Rogers produced a theory of the adoption of innovations among individuals and organizations. *Diffusion of Innovations* and Rogers' later books are among the most often cited in diffusion research. His methodologies are closely followed in recent diffusion research, even as the field has expanded into, and been influenced by, other methodological disciplines such as social network analysis and communication.

Elements

The key elements in diffusion research are:

Element	Definition
Innovation	Innovations are a broad category, relative to the current knowledge of the analyzed unit. Any idea, practice, or object that is perceived as new by an individual or other unit of adoption could be considered an innovation available for study.
Adopters	Adopters are the minimal unit of analysis. In most studies, adopters are individuals, but can also be organizations (businesses, schools, hospitals, etc.), clusters within social networks, or countries.
Communication channels	Diffusion, by definition, takes place among people or organizations. Communication channels allow the transfer of information from one unit to the other. Communication patterns or capabilities must be established between parties as a minimum for diffusion to occur.
Time	The passage of time is necessary for innovations to be adopted; they are rarely adopted instantaneously. In fact, in the Ryan and Gross (1943) study on hybrid corn adoption, adoption occurred over more than ten years, and most farmers only dedicated a fraction on their fields to the new corn in the first years after adoption.
Social system	The social system is the combination of external influences (mass media, organizational or governmental mandates) and internal influences (strong and weak social relationships, distance from opinion leaders). There are many roles in a social system, and their combination represents the total influences on a potential adopter.

Characteristics of Innovations

Studies have explored many characteristics of innovations. Meta-reviews have identified several characteristics that are common among most studies. These are in line with the characteristics that Rogers initially cited in his reviews.

Potential adopters evaluate an innovation on its relative advantage (the perceived efficiencies gained by the innovation relative to current tools or procedures), its compatibility with the pre-existing system, its complexity or difficulty to learn, its trialability or testability, its potential for reinvention (using the tool for initially unintended purposes), and its observed effects. These qualities interact and are judged as a whole. For example, an innovation might be extremely complex, reducing its likelihood to be adopted and diffused, but it might be very compatible with a large advantage relative to current tools. Even with this high learning curve, potential adopters might adopt the innovation anyway.

Studies also identify other characteristics of innovations, but these are not as common as the ones that Rogers lists above. The fuzziness of the boundaries of the innovation can impact its adoption. Specifically, innovations with a small core and large periphery are easier to adopt. Innovations that are less risky are easier to adopt as the potential loss from failed integration is lower. Innovations that are disruptive to routine tasks, even when they bring a large relative advantage, might not be adopted because of added instability. Likewise, innovations that make tasks easier are likely to be adopted. Closely related to relative complexity, knowledge requirements are the ability barrier to use presented by the difficulty to use the innovation. Even when there are high knowledge requirements, support from prior adopters or other sources can increase the chances for adoption.

Characteristics of Individual Adopters

Like innovations, adopters have been determined to have traits that affect their likelihood to adopt an innovation. A bevy of individual personality traits have been explored for their impacts on adoption, but with little agreement. Ability and motivation, which vary on situation unlike personality traits, have a large impact on a potential adopter's likelihood to adopt an innovation. Unsurprisingly, potential adopters who are motivated to adopt an innovation are likely to make the adjustments needed to adopt it. Motivation can be impacted by the meaning that an innovation holds; innovations can have symbolic value that encourage (or discourage) adoption. First proposed by Ryan and Gross (1943), the overall connectedness of a potential adopter to the broad community represented by a city. Potential adopters who frequent metropolitan areas are more likely to adopt an innovation. Finally, potential adopters who have the power or agency to create change, particularly in organizations, are more likely to adopt an innovation than someone with less power over his choices.

Characteristics of Organizations

Organizations face more complex adoption possibilities because organizations are both the aggregate of its individuals and its own system with a set of procedures and norms. Three organizational characteristics match well with the individual characteristics above: tension for change (motivation and ability), innovation-system fit (compatibility), and assessment of implications (observability). Organizations can feel pressured by a tension for change. If the organization's situation is untenable, it will be motivated to adopt an innovation to change its fortunes. This tension often plays out among its individual members. Innovations that match the organization's pre-existing system require fewer coincidental changes and are easy to assess are more likely to be adopted. The wider environment of the organization, often an industry, community, or economy, exerts pressures on the organization, too. Where an innovation is diffusing through the organization's environment for any reason, the organization is more likely to adopt it. Innovations that are intentionally spread, including by political mandate or directive, are also likely to diffuse quickly.

Process

Diffusion occurs through a five–step decision-making process. It occurs through a series of communication channels over a period of time among the members of a similar social system. Ryan and Gross first identified adoption as a process in 1943. Rogers' five stages (steps): awareness, interest, evaluation, trial, and adoption are integral to this theory. An individual might reject an innovation at any time during or after the adoption process. Abrahamson examined this process critically by posing questions such as: How do technically inefficient innovations diffuse and what impedes technically efficient innovations from catching on? Abrahamson makes suggestions for how organizational scientists can more comprehensively evaluate the spread of innovations. In later editions of *Diffusion of Innovation*, Rogers changes his terminology of the five stages to: knowledge, persuasion, decision, implementation, and confirmation. However, the descriptions of the categories have remained similar throughout the editions.

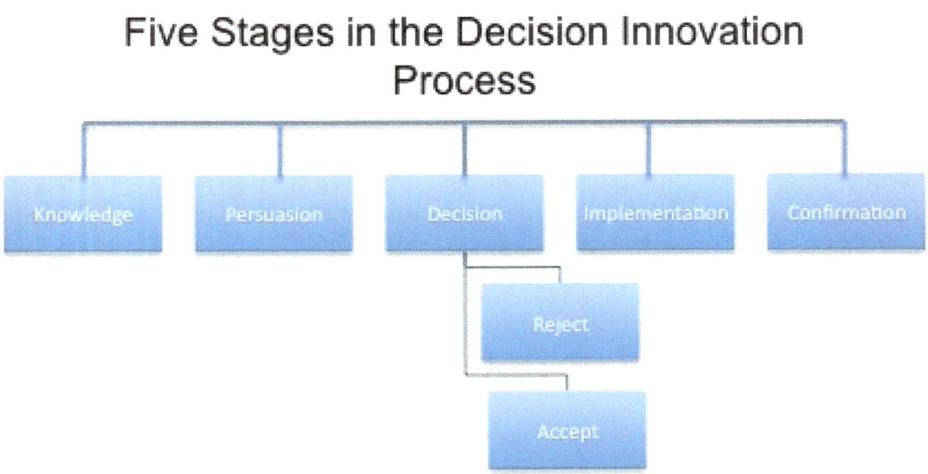

Five stages of the adoption process	
Stage	**Definition**
Knowledge	The individual is first exposed to an innovation, but lacks information about the innovation. During this stage the individual has not yet been inspired to find out more information about the innovation.
Persuasion	The individual is interested in the innovation and actively seeks related information/details.
Decision	The individual takes the concept of the change and weighs the advantages/disadvantages of using the innovation and decides whether to adopt or reject the innovation. Due to the individualistic nature of this stage, Rogers notes that it is the most difficult stage on which to acquire empirical evidence.
Implementation	The individual employs the innovation to a varying degree depending on the situation. During this stage the individual also determines the usefulness of the innovation and may search for further information about it.
Confirmation	The individual finalizes his/her decision to continue using the innovation. This stage is both intrapersonal (may cause cognitive dissonance) and interpersonal, confirmation the group has made the right decision.

Decisions

Two factors determine what type a particular decision is:

- Whether the decision is made freely and implemented voluntarily
- Who makes the decision.

Based on these considerations, three types of innovation-decisions have been identified.

Type	**Definition**
Optional Innovation-Decision	made by an individual who is in some way distinguished from others.
Collective Innovation-Decision	made collectively by all participants.
Authority Innovation-Decision	made for the entire social system by individuals in positions of influence or power.

Rate of Adoption

The rate of adoption is defined as the relative speed at which participants adopt an innovation. Rate is usually measured by the length of time required for a certain percentage of the members of a social system to adopt an innovation. The rates of adoption for innovations are determined by an individual's adopter category. In general, individuals who first adopt an innovation require a shorter adoption period (adoption process) when compared to late adopters.

Within the adoption curve at some point the innovation reaches critical mass. This is when the number of individual adopters ensures that the innovation is self-sustaining.

Adoption Strategies

Rogers outlines several strategies in order to help an innovation reach this stage, including when an innovation adopted by a highly respected individual within a social network and creating an instinctive desire for a specific innovation. Another strategy includes injecting an innovation into a group of individuals who would readily use said technology, as well as providing positive reactions and benefits for early adopters.

Diffusion vs Adoption

Adoption is an individual process detailing the series of stages one undergoes from first hearing about a product to finally adopting it. Diffusion signifies a group phenomenon, which suggests how an innovation spreads.

Adopter Categories

Rogers defines an adopter category as a classification of individuals within a social system on the basis of innovativeness. In the book *Diffusion of Innovations*, Rogers suggests a total of five categories of adopters in order to standardize the usage of adopter categories in diffusion research. The adoption of an innovation follows an S curve when plotted over a length of time. The categories of adopters are: innovators, early adopters, early majority, late majority and laggards In addition to the gatekeepers and opinion leaders who exist within a given community, change agents may come from outside the community. Change agents bring innovations to new communities– first through the gatekeepers, then through the opinion leaders, and so on through the community.

Adopter category	**Definition**
Innovators	Innovators are willing to take risks, have the highest social status, have financial liquidity, are social and have closest contact to scientific sources and interaction with other innovators. Their risk tolerance allows them to adopt technologies that may ultimately fail. Financial resources help absorb these failures.
Early adopters	These individuals have the highest degree of opinion leadership among the adopter categories. Early adopters have a higher social status, financial liquidity, advanced education and are more socially forward than late adopters. They are more discreet in adoption choices than innovators. They use judicious choice of adoption to help them maintain a central communication position.

Early Majority	They adopt an innovation after a varying degree of time that is significantly longer than the innovators and early adopters. Early Majority have above average social status, contact with early adopters and seldom hold positions of opinion leadership in a system (Rogers 1962, p. 283)
Late Majority	They adopt an innovation after the average participant. These individuals approach an innovation with a high degree of skepticism and after the majority of society has adopted the innovation. Late Majority are typically skeptical about an innovation, have below average social status, little financial liquidity, in contact with others in late majority and early majority and little opinion leadership.
Laggards	They are the last to adopt an innovation. Unlike some of the previous categories, individuals in this category show little to no opinion leadership. These individuals typically have an aversion to change-agents. Laggards typically tend to be focused on "traditions", lowest social status, lowest financial liquidity, oldest among adopters, and in contact with only family and close friends.

Failed Diffusion

Failed diffusion does not mean that the technology was adopted by no one. Rather, failed diffusion often refers to diffusion that does not reach or approach 100% adoption due to its own weaknesses, competition from other innovations, or simply a lack of awareness. From a social networks perspective, a failed diffusion might be widely adopted within certain clusters but fail to make an impact on more distantly related people. Networks that are over-connected might suffer from a rigidity that prevents the changes an innovation might bring, as well. Sometimes, some innovations also fail as a result of lack of local involvement and community participation.

For example, Rogers discussed a situation in Peru involving the implementation of boiling drinking water to improve health and wellness levels in the village of Los Molinas. The residents had no knowledge of the link between sanitation and illness. The campaign worked with the villagers to try to teach them to boil water, burn their garbage, install latrines and report cases of illness to local health agencies. In Los Molinas, a stigma was linked to boiled water as something that only the "unwell" consumed, and thus, the idea of healthy residents boiling water prior to consumption was frowned upon. The two-year educational campaign was considered to be largely unsuccessful. This failure exemplified the importance of the roles of the communication channels that are involved in such a campaign for social change. An examination of diffusion in El Salvador determined that there can be more than one social network at play as innovations are communicated. One network carries information and the other carries influence. While people might hear of an innovation's uses, in Rogers' Los Molinas sanitation case, a network of influence and status prevented adoption.

Heterophily and Communication Channels

Lazarsfeld and Merton first called attention to the principles of homophily and its opposite, heterophily. Using their definition, Rogers defines homophily as "the degree to which pairs of individuals who interact are similar in certain attributes, such as beliefs, education, social status, and the like". When given the choice, individuals usually choose to interact with someone similar to themselves. Homophilous individuals engage in more effective communication because their similarities lead to greater knowledge gain as well as attitude or behavior change. As a result, homophilous people tend to promote diffusion among each other. However, diffusion requires a certain degree of heterophily to introduce new ideas into a relationship; if two individuals are identical, no diffusion occurs because there is no new information to exchange. Therefore, an ideal situation would involve potential adopters who are homophilous in every way, except in knowledge of the innovation.

Promotion of healthy behavior provides an example of the balance required of homophily and heterophily. People tend to be close to others of similar health status. As a result, people with unhealthy behaviors like smoking and obesity are less likely to encounter information and behaviors that encourage good health. This presents a critical challenge for health communications, as ties between heterophilous people are relatively weaker, harder to create, and harder to maintain. Developing heterophilous ties to unhealthy communities can increase the effectiveness of the diffusion of good health behaviors. Once one previously homophilous tie adopts the behavior or innovation, the other members of that group are more likely to adopt it, too.

The Role of Social Systems

Opinion Leaders

Not all individuals exert an equal amount of influence over others. In this sense opinion leaders are influential in spreading either positive or negative information about an innovation. Rogers relies on the ideas of Katz & Lazarsfeld and the two-step flow theory in developing his ideas on the influence of opinion leaders.

Opinion leaders have the most influence during the evaluation stage of the innovation-decision process and on late adopters. In addition opinion leaders typically have greater exposure to the mass media, more cosmopolitan, greater contact with change agents, more social experience and exposure, higher socioeconomic status, and are more innovative than others.

Research was done in the early 1950s at the University of Chicago attempting to assess the cost-effectiveness of broadcast advertising on the diffusion of new products and services. The findings were that opinion leadership tended to be organized into a hierarchy within a society, with each level in the hierarchy having most influence over other members in the same level, and on those in the next level below it. The lowest levels were generally larger in numbers and tended to coincide with various demographic attributes that might be targeted by mass advertising. However, it found that direct word of mouth and example were far more influential than broadcast messages, which were only effective if they reinforced the direct influences. This led to the conclusion that advertising was best targeted, if possible, on those next in line to adopt, and not on those not yet reached by the chain of influence.

Other research relating the concept to public choice theory finds that the hierarchy of influence for innovations need not, and likely does not, coincide with hierarchies of official, political, or economic status. Elites are often not innovators, and innovations may have to be introduced by outsiders and propagated up a hierarchy to the top decision makers.

Electronic Communication Social Networks

Prior to the introduction of the Internet, it was argued that social networks had a crucial role in the diffusion of innovation particularly tacit knowledge in the book *The IRG Solution – hierarchical incompetence and how to overcome it*. The book argued that the widespread adoption of computer networks of individuals would lead to much better diffusion of innovations, with greater understanding of their possible shortcomings and the identification of needed innovations that would not have otherwise occurred. The social model proposed by Ryan and Gross is expanded by Valente who uses social networks as a basis for adopter categorization instead of solely relying on the system-level analysis used by Ryan and Gross. Valente also looks at an individual's personal network, which is a different application than the organizational perspective espoused by many other scholars.

Recent research by Wear shows, that particularly in regional and rural areas, significantly more innovation takes place in communities which have stronger inter-personal networks.

Organizations

Innovations are often adopted by organizations through two types of innovation-decisions: collective innovation decisions and authority innovation decisions. The collective decision occurs when adoption is by consensus. The authority decision occurs by adoption among very few individuals with high positions of power within an organization. Unlike the optional innovation decision process, these decision processes only occur within an organization or hierarchical group. Within an organization certain individuals are termed "champions" who stand behind an innovation and break through opposition. The champion plays a very similar role as the champion used within the efficiency business model Six Sigma. The process contains five stages that are slightly similar to the innovation-decision process that individuals undertake. These stages are: agenda-setting, matching, redefining/restructuring, clarifying and routinizing.

Extensions of the Theory

Policy

Diffusion of Innovations has been applied beyond its original domains. In the case of political science and administration, policy diffusion focuses on how institutional innovations are adopted by other institutions, at the local, state, or country level. An alternative term is 'policy transfer' where the focus is more on the agents of diffusion and the diffusion of policy knowledge, such as in the work of Diane Stone. Specifically, policy transfer can be defined as "knowledge about how policies administrative arrangements, institutions, and ideas in one political setting (past or present) is used in the development of policies, administrative arrangements, institutions, and ideas in another political setting".

The first interests with regards to policy diffusion were focused in time variation or state lottery adoption, but more recently interest has shifted towards mechanisms (emulation, learning and coercion) or in channels of diffusion where researchers find that regulatory agency creation is transmitted by country and sector channels. At the local level, examining popular city-level policies make it easy to find patterns in diffusion through measuring public awareness. At the international level, economic policies have been thought to transfer among countries according to local politicians' learning of successes and failures elsewhere and outside mandates made by global financial organizations. As a group of countries succeed with a set of policies, others follow, as exemplified by the deregulation and liberalization across the developing world after the successes of the Asian Tigers. The reintroduction of regulations in the early 2000s also shows this learning process, which would fit under the stages of knowledge and decision, can be seen as lessons learned by following China's successful growth.

Technology

Peres, Muller and Mahajan suggested that diffusion is "the process of the market penetration of new products and services that is driven by social influences, which include all interdependencies among consumers that affect various market players with or without their explicit knowledge".

Eveland evaluated diffusion from a phenomenological view, stating, "Technology is information, and exists only to the degree that people can put it into practice and use it to achieve values".

Diffusion of existing technologies has been measured using "S curves". These technologies include radio, television, VCR, cable, flush toilet, clothes washer, refrigerator, home ownership, air conditioning, dishwasher, electrified households, telephone, cordless phone, cellular phone, per capita airline miles, personal computer and the Internet. These data can act as a predictor for future innovations.

Diffusion curves for infrastructure reveal contrasts in the diffusion process of personal technologies versus infrastructure.

Consequences of Adoption

Both positive and negative outcomes are possible when an individual or organization chooses to adopt a particular innovation. Rogers states that this area needs further research because of the biased positive attitude that is associated with innovation. Rogers lists three categories for consequences: desirable vs. undesirable, direct vs. indirect, and anticipated vs. unanticipated.

In contrast Wejnert details two categories: public vs. private and benefits vs. costs.

Public versus Private

Public consequences comprise the impact of an innovation on those other than the actor, while private consequences refer to the impact on the actor. Public consequences usually involve collective actors, such as countries, states, organizations or social movements. The results are usually concerned with issues of societal well-being. Private consequences usually involve individuals or small collective entities, such as a community. The innovations are usually concerned with the improvement of quality of life or the reform of organizational or social structures.

Benefits versus Costs

Benefits of an innovation obviously are the positive consequences, while the costs are the negative. Costs may be monetary or nonmonetary, direct or indirect. Direct costs are usually related to financial uncertainty and the economic state of the actor. Indirect costs are more difficult to identify. An example would be the need to buy a new kind of pesticide to use innovative seeds. Indirect costs may also be social, such as social conflict caused by innovation. Marketers are particularly interested in the diffusion process as it determines the success or failure of a new product. It is quite important for a marketer to understand the diffusion process so as to ensure proper management of the spread of a new product or service.

Mathematical Treatment

The diffusion of an innovation typically follows an S shaped curve which often resembles a logistic function. Mathematical programming models such as the S-D model apply the diffusion of innovations theory to real data problems.

Complex Systems Models

Complex network models can also be used to investigate the spread of innovations among individuals connected to each other by a network of peer-to-peer influences, such as in a physical community or neighborhood.

Such models represent a system of individuals as *nodes* in a network (or graph). The interactions that link these individuals are represented by the edges of the network and can be based on the probability or strength of social connections. In the dynamics of such models, each node is assigned a current state, indicating whether or not the individual has adopted the innovation, and model equations describe the evolution of these states over time.

In threshold models, the uptake of technologies is determined by the balance of two factors: the (perceived) usefulness (sometimes called utility) of the innovation to the individual as well as barriers to adoption, such as cost. The multiple parameters that influence decisions to adopt, both individual and socially motivated, can be represented by such models as a series of nodes and connections that represent real relationships. Borrowing from social network analysis, each node is an innovator, an adopter, or a potential adopter. Potential adopters have a threshold, which is a fraction of his neighbors who adopt the innovation that must be reached before he will adopt. Over time, each potential adopter views his neighbors and decides whether he should adopt based on the technologies they are using. When the effect of each individual node is analyzed along with its influence over the entire network, the expected level of adoption was seen to depend on the number of initial adopters and the network's structure and properties. Two factors emerge as important to successful spread of the innovation: the number of connections of nodes with their neighbors and the presence of a high degree of common connections in the network (quantified by the clustering coefficient). These models are particularly good at showing the impact of opinion leaders relative to others. Computer models are often used to investigate this balance between the social aspects of diffusion and perceived intrinsic benefit to the individuals.

Criticism

Because there are more than four thousand articles across many disciplines published on Diffusion of Innovations, with a vast majority written after Rogers created a systematic theory, there have been few widely adopted changes to the theory. Although each study applies the theory in slightly different ways, this lack of cohesion has left the theory stagnant and difficult to apply with consistency to new problems.

Diffusion is difficult to quantify because humans and human networks are complex. It is extremely difficult, if not impossible, to measure what exactly causes adoption of an innovation. This is important, particularly in healthcare. Those encouraging adoption of health behaviors or new medical technologies need to be aware of the many forces acting on an individual and his or her decision to adopt a new behavior or technology. Diffusion theories can never account for all variables, and therefore might miss critical predictors of adoption. This variety of variables has also led to inconsistent results in research, reducing heuristic value.

Rogers placed the contributions and criticisms of diffusion research into four categories: pro-innovation bias, individual-blame bias, recall problem, and issues of equality. The pro-innovation bias, in particular, implies that all innovation is positive and that all innovations should be adopted. Cultural traditions and beliefs can be consumed by another culture's through diffusion, which can impose significant costs on a group of people. The one-way information flow, from sender to receiver, is another weakness of this theory. The message sender has a goal to persuade the receiver, and there is little to no reverse flow. The person implementing the change controls the direction and outcome of the campaign. In some cases, this is the best approach, but other cases require a more participatory approach. In complex environments where the adopter is receiving information from many sources and is returning feedback to the sender, a one-way model is insufficient and multiple communication flows need to be examined.

Innovation Economics

Innovation economics is a growing economic theory that emphasizes entrepreneurship and innovation. Innovation economics is based on two fundamental tenets: that the central goal of economic policy should be to spur higher productivity through greater innovation, and that markets relying on input resources and price signals alone will not always be as effective in spurring higher productivity, and thereby economic growth.

This is in contrast to the two other conventional economic doctrines, neoclassical economics and Keynesian economics.

Historical Origins

Joseph Schumpeter was one of the first and most important scholars who extensively has tackle the question of innovation in Economics. In contrast to his contemporary John Maynard Keynes, Schumpeter contended that evolving institutions, entrepreneurs, and technological change were at the heart of economic growth, not independent forces that are largely unaffected by policy. He argued that "capitalism can only be understood as an evolutionary process of continuous innovation and 'creative destruction".

But it is only within the last 15 years that a theory and narrative of economic growth focused on innovation that was grounded in Schumpeter's ideas has emerged. Innovation economics attempted to answer the fundamental problem in the puzzle of total factor productivity growth. Continual growth of output could no longer be explained only in increase of inputs used in the production process as understood in industrialization. Hence, innovation economics focused on a theory of economic creativity that would impact the theory of the firm and organization decision-making. Hovering between heterodox economics that emphasized the fragility of conventional assumptions and orthodox economics that ignored the fragility of such assumptions, innovation economics aims for joint didactics between the two. As such, it enlarges the Schumpeterian analyses of new technological system by incorporating new ideas of information and communication technology in the global economy.

Indeed, a new theory and narrative of economic growth focused on innovation has emerged in the last decade. Innovation economics emerges on the wage of other schools of thoughts in economics, including new institutional economics, new growth theory, endogenous growth theory, evolutionary economics, neo-Schumpeterian economics—provides an economic framework that explains and helps support growth in today's knowledge economy.

Leading theorists of innovation economics include both formal economists, as well as management theorists, technology policy experts, and others. These include Paul Romer, Elhanan Helpman, W. Brian Arthur, Robert Axtell, Richard R. Nelson, Richard Lipsey, Michael Porter, Christopher Freeman.

Theory

Innovation economists believe that what primarily drives economic growth in today's knowledge-based economy is not capital accumulation, as claimed by neoclassicalism asserts, but innovative capacity spurred by appropriable knowledge and technological externalities. Economics growth in innovation economics is the end-product of knowledge (tacit vs. codified); regimes and policies allowing for entrepreneurship and innovation (i.e., R&D expenditures, permits, licenses); technological spillovers and externalities between collaborative firms; and systems of innovation that create innovative environments (i.e., clusters, agglomerations, metropolitan areas).

In 1970, economist Milton Friedman said in the *New York Times* that a business's sole purpose is to generate profits for their shareholders and companies that pursued other missions would be less competitive, resulting in fewer benefits to owners, employees, and society. Yet data over the past several decades shows that while profits matter, good firms supply far more, particularly in bringing innovation to the market. This fosters economic growth, employment gains, and other society-wide benefits. Business school professor David Ahlstrom asserts: "the main goal of business is to develop new and innovative goods and services that generate economic growth while delivering benefits to society."

In contrast to neoclassical economics, innovation economics offer differing perspectives on main focus, reasons for economic growth, and the assumptions of context between economic actors:

Economic thought	Focus	Growth	Context
Neoclassical	Market price signals in using scarce resources	Productive factor accumulation (capital, labor)	Individuals and firms behaving in vacuum

Innovation	Innovative capacity to create more effective processes, products, business models	Knowledge/technology (R&D, patents)	Institutions of research, government, society

Despite the differences in economic thought, both perspectives are based on the same core premise: the foundation of all economic growth is the optimization of the utilization of factors and the measure of success is how well the factor utilization is optimized. Whatever the factors, it nonetheless leads to the same situation of special endowments, varying relative prices, and production processes. So while, the two differ in theoretical concepts, innovation economics can find fertile ground in mainstream economics, rather than remain in diametric contention.

Evidence

Empirical evidence worldwide points to a positive link between technological innovation and economic performance. The drive of biotech firms in Germany was due to the R&D subsidies to joint projects, network partners, and close cognitive distance of collaborative partners within a cluster. These factors increased patent performance in the biotech industry. Additionally, innovation capacity explains much of the GDP growth in India and China between 1981–2004 but especially in the 1990s. Their development of a National Innovation System through heavy investment of R&D expenditures and personnel, patents, and high-tech/service exports strengthened their innovation capacity. By linking the science sector with the business sector, establishing incentives for innovative activities, and balancing the import of technology and indigenous R&D effort, both countries experienced rapid economic growth in recent decades. Also, the Council of Foreign Relations asserted that since the end of the 1970s, the U.S. has gained a disproportionate share of the world's wealth through their aggressive pursuit of technological change, demonstrating that technological innovation is a central catalyst of steady economic performance. Concisely, evidence shows that innovation contributes to steady economic growth and rise in per capita income.

However, some empirical studies investigating the innovation-performance-link lead to rather mixed results and indicate that the relationship be more subtle and complex than commonly assumed. In particular, the relationship between innovativeness and performance seems to differ in intensity and significance across empirical contexts, environmental circumstances, and conceptual dimensions.

All of the above has taken place in an era of data constraint, as identified by Zvi Griliches twenty years ago. Because the primary domain of innovation is commerce the key data resides there; continually out of campus reach in reports hidden within factories, corporate offices and technical centers. This recusal still stymies progress today. Recent attempts at data transference have led, not least, to the 'positive link' (above) being upgraded to exact algebra between R&D productivity and GDP allowing prediction from one to the other. This is pending further disclosure from commercial sources but several pertinent documents are already available.

Geography

While innovation is important, it is not a happenstance occurrence as a natural harbor or natural resources are, but a deliberate, concerted effort of markets, institutions, policymakers, and effect use of geographic space. In global economic restructuring, location has become a key element in establishing competitive advantage as regions focus on their unique assets to spur innovation (i.e., information technology in Silicon Valley, CA; digital media in Seoul, South Korea). Even more, thriving metropolitan economies that carry multiple clusters (i.e., Tokyo, Chicago, London) essentially fuel national economies through their pools of human capital, innovation, quality places, and infrastructure. Cities become "innovative spaces" and "cradles of creativity" as drivers of innovation. They become essential to the system of innovation through the supply side: ready, available, abundant capital and labor; good infrastructure for productive activities, and diversified production structures that spawn synergies and hence innovation. In addition they grow due to the demand side: diverse population of varying occupations, ideas, skills; high and differentiated level of consumer demand; and constant recreation of urban order especially infrastructure of streets, water systems, energy, and transportation.

Induced Innovation

Induced innovation is a macroeconomic hypothesis first proposed in 1932 by J.R. Hicks in his work *The Theory of Wages*. He proposed that "a change in the relative prices of the factors of production is itself a spur to invention, and to invention of a particular kind—directed to economizing the use of a factor which has become relatively expensive."

Considerable literature has been produced on this hypothesis, which is often presented in terms of the effects of wage increases as an encouragement to labor-saving innovation. The hypothesis has also been applied to viewing increases in energy costs as a motivation for a more rapid improvement in energy efficiency of goods than would normally occur.

Induced Innovation in Climate Change

A significant application of Hicks theory can be found in the field of climate change. The exponential population growth occurred in the last century has drastically increased pressure on natural resources. In order to have a sustainable future it's imperative to modify global strategies on climate change and the induced innovation theory can aid to model these policies.

To calculate the human impact on the environment economists often use the I=P*A*T equation where "I", the impact variable, (for example energy consumption) is the product of "P", the population, "A" the affluence (often embodied by GDPper capita) and "T" the technology.

The technical coefficient represents the efficiency of the system in use for particular resource and expresses the average state of technology. The decrease of "T" would indicate a gain in efficiency however "I" could still be growing or remaining stable if the improved technology is not sufficient to compensate the effect of an increase in "P" and "A". Therefore, a reduction in "I" would always mean that pressure on resources has lightened but this might not always be the consequence of using resources more efficiently (reducing T).

A fundamental issue of climate change is the excess of greenhouse gasses in the atmosphere particularly CO_2 as the result of an intensified economic activity. Global GDP and CO_2 emissions were growing at a corresponding rate in the until the 1970s. It was then that oil prices have increased exponentially causing people to reduce its consumption.

According to Hicks theory a change in factor prices will stimulate development of technical innovation to reduce the use of the factor for which the price increased compared to other factors. Following the oil shock significant investments were made in alternative sources of energy, more efficient cars and heating systems to mitigate oil consumption. As a result, CO_2 emissions started growing at a slower rate than GDP per capita. Although petrol prices then dropped in the 1980s, CO_2 emissions have continued to grow more slowly than GDP. This is an indication of a complete structural change in technology induced by the need to innovate.

Induced Innovation in Agriculture

The development of agriculture is a fundamental part of the world economic growth. Being able to meet the demand of an exponentially growing population can be a challenging task, especially if we consider that one of the most important factors, land, remains fairly fixed.

According to the World Bank data, less than 38% of the global surface area of a can be deemed usable for cultivation of crops or permanent pastures.

Expanding such a number can only be achieved by remediation of soil in a certain area in order to make the land fertile. Needless to say that such an intervention requires a huge investment that can be amortized only in a long period of time. Therefore, to meet the increasing demand, countries are forced to drastically increase the productivity of the land they dispose. This is where induced innovation steps in.

In order to best explain how Hick's macroeconomic theory works, let us consider a situation in which the demand for agricultural derived products increases as a result of either population growth or increased household income. Theory states that in such a situation, prices of inputs for which supply is inelastic will rise relative to prices of more elastic inputs. Similarly, if the supply of a particular input increases at a faster rate than the supply of other inputs, the price of such input will decline relative to the price of the other factors of production used. Ideally then, farmers would be looking to replace or use less of the more inelastic and less responsive factors of production since they are the more expensive to use. Therefore, technical innovations that replace such inputs would guarantee less costs and hence more profits. In other words, when demand for their products increases, farmers are lured by changes in relative prices to seek for technological alternatives that substitute the increasingly scarce factors of production.

Here is where the government has to intervene. Clearly, agricultural workers can't develop these innovations themselves and therefore demand that public research institutions develop new technologies that will then be transformed into modernized machinery that agricultural supply firms can sell to local farmers. Scientists will now study the best solution to this problem and respond by making the appropriate technical development such that producers can efficiently substitute the abundant inputs for the increasingly scarce factors. Ultimately this will lead to a reduction of farmer's unit costs in the best possible way.

Criticism to the Induced Innovation Theory

Although some validity is still today attributed to John Hicks induced innovation theory, different economists have critiqued his claims and developed the theory further.

There has been a massive difference of technical innovation between developed and underdeveloped countries. LDC's are expected, in the imminent future, to industrialize and develop new technologies. Economists believe that induced innovation can have a significant role in rural farming technical advancement. However Danish economist Ester Boserup takes a slightly different position. She believes that population growth would increase pressure on soil leading to improved soil management strategies aimed to raise productivity. Contrarily to Hicks that would claim this is the consequence of an increase in factor prices, Boserup believes growing population number and physical land availability are the main determinants of this change.

Similarly economist Skarstein takes a detached position from Hicks theory. He argues that the induced innovation can't be an automatic outcome of an increase in factor prices. "Shifting factor prices would not lead to any innovations in factor saving techniques, it will merely induce a change in techniques along the prevailing substitution line, he argues. This change will take place as soon as the relative factor prices shift. When the change has taken place, there is no further incentive to move inwards to a more efficient technology."

Salters sustained another contrasting view from Hicks theory. He claimed that all firms are cost savers; therefore under competitive equilibrium regime factors of production are paid its marginal value. Consequently, they will all have the same value leading to no particular stimulus to innovate to save a specific factor.

Linear Model of Innovation

The Linear Model of Innovation is an early model of innovation that suggests technical change happens in a linear fashion from Invention to Innovation to Diffusion.

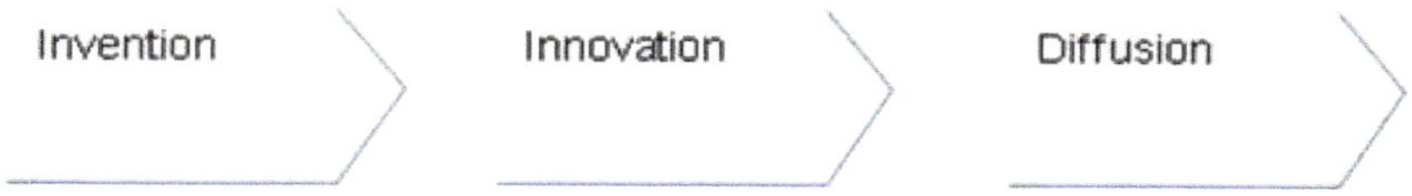

Original model of three phases of the process of Technological Change

It prioritises scientific research as the basis of innovation, and plays down the role of later players in the innovation process.

Current models of innovation deriving from approaches such as Actor-Network Theory, Social shaping of technology and social learning (e.g. Tuomi, (2002)),provide a much richer picture of the way innovation works.

Current ideas in Open Innovation and User innovation derive from these later ideas.

Also known as 'Traditional Phase Gate Model', under this model, product or services concept is frozen at early stage so as to minimize risk. Also innovation process in enterprise involves series of sequential phases/steps arranged in such a manner that the preceding phase must be cleared before moving to next phase. Thus a project must pass through a gate with the permission of gatekeeper before moving to the next succeeding phase.

Criteria for passing through each gate, and the person at each gate are defined beforehand. The gatekeeper examines whether the stated objectives for preceding phase have been properly met or not and whether desired development has taken place at the preceding phase or not.

Two versions of the linear model of innovation are often presented:

1. "technology push" model
2. "market pull" model.

From the 1950s to the Mid-1960s, the industrial innovation process was generally perceived as a linear progression from scientific discovery, through technological development in firms, to the marketplace, Rothwell (1994) The stages of the "Technology Push" model are:

Basic science→Design and engineering→Manufacturing→Marketing→Sales

From the Mid 1960s to the Early 1970s, emerges the second-generation Innovation model, referred to as the "market pull" model of innovation. According to this simple sequential model, the market was the source of new ideas for directing R&D, which had a reactive role in the process. The stages of the "market pull " model are:

Market need—Development—Manufacturing—Sales.

The linear models of innovation supported numerous criticisms concerning the linearity of the models. These models ignore the many feedbacks and loops that occur between the different "stages" of the process. Shortcomings and failures that occur at various stages may lead to a reconsideration of earlier steps and this may result in an innovation. A history of the linear model of innovation may be found in Godin (2006)

Social Theory

Social theories are frameworks of empirical evidence used to study and interpret social phenomena. A tool used by social scientists, social theories relate to historical debates over the most valid and reliable methodologies (e.g. positivism and antipositivism), as well as the primacy of either structure or agency. Certain social theories attempt to remain strictly scientific, descriptive, and objective. Conflict theories, by contrast, present ostensibly normative positions, and often critique the ideological aspects inherent in conventional, traditional thought.

The origins of social theory are difficult to pinpoint, but debates frequently return to ancient Greece (Berberoglu 2005, p. xi). From these foundations in Western philosophy arose Enlightenment social contract theory, sociological positivism, and modern social science. Today, 'social science' is used as an umbrella term to refer to sociology, economics, political science, jurisprudence, and other disciplines. Social theory is interdisciplinary and draws upon ideas from fields as diverse as anthropology and media studies. Social theory of an informal nature, or authorship based outside of academic social and political science, may be referred to instead as "social criticism" or "social commentary". Similarly, "cultural criticism" may be associated both with formal cultural and literary scholarship, as well as other non-academic or journalistic forms of writing.

Social theory as a distinct discipline emerged in the 20th century and was largely equated with an attitude of critical thinking, based on rationality, logic and objectivity, and the desire for knowledge through *a posteriori* methods of discovery, rather than *a priori* methods of tradition.

Social Thought

Social thought provides general theories to explain actions and behavior of society as a whole, encompassing sociological, political, and philosophical ideas. Social theory is used to make distinctions and generalizations among different types of societies, and to analyze modernity as it has emerged in the past few centuries. Classical social theory has generally been presented from a perspective of Western philosophy, and often regarded as Eurocentric.

History

Greek Philosophy

Ancient Greek philosophers, including Aristotle and Plato, did not see a distinction between politics and society. The concept of society did not come until much later, during the Enlightenment period. The term, *société*, was probably first used as key concept by Rousseau in discussion of social relations.

Pre-enlightenment Social Theory

The earliest proto-social scientific observations are to be found in the founding texts of Western philosophy (Herodotus, Thucydides, Plato, Polybius and so on), as well as in the non-European thought of figures such as Confucius. Prior to the enlightenment, social theory took largely narrative and normative form. Expressed as stories and fables, it may be assumed the pre-Socratic philosophers and religious teachers were the precursors to social theory proper.

Saint Augustine (354–430) and St. Thomas Aquinas (*circa* 1225–1274) concerned themselves exclusively with the idea of the *just society*. St. Augustine describes late Ancient Roman society but through a lens of hatred and contempt for what he saw as false Gods, and in reaction theorized City of God. Similarly, in China, Master Kong (otherwise known as Confucius) (551–479 BCE) envisaged a just society that went beyond his contemporary society of the Warring States. Later on, also in China, Mozi (*circa* 470 – *circa* 390 BCE) recommended a more pragmatic sociology, but ethical at base.

Sociology in Medieval Islam

There is evidence of early Muslim sociology from the 14th century: Ibn Khaldun, in his *Muqaddimah* (later translated as *Prolegomena* in Latin), the introduction to a seven volume analysis of universal history, was the first to advance social philosophy and social science in formulating theories of social cohesion and social conflict. He is thus considered by many to be the forerunner of sociology.

Ibn Khaldun

Ibn Khaldun, an influential Muslim scholar, described in *Muqaddimah (Introduction to History)*, published in 1377, two types of societies: (1) the city or town-dweller and (2) the mobile, nomadic societies.

Montesquieu

Montesquieu, in *The Spirit of Laws*, was possibly the first to suggest a universal explanation for history. Another innovative aspect of Montesquieu's thinking was that he included changes in mores and manners as part of his explanation of political and historic events.

Political Philosophy and Social Contract Theory

During the Age of Enlightenment, political entities expanded from basic systems of self-governance and monarchy to the complex democratic and communist systems that exist in the Industrialized and the Modern Eras. In the 18th century, after Montesquieu's *The Spirit of the Laws* established that social elements influence human nature, the pre-classical period of social theories developed a new form that provides the basic ideas for social theory, such as: evolution, philosophy of history, social life and social contract, public and general will, competition in social space, organistic pattern for social description and so forth. Jean-Jacques Rousseau in this time played a significant role in social theory. He revealed the origin of inequality, analyzed the social contract (and social compact) that forms social integration and defined the social sphere or civil society. He also emphasized that man has the liberty to change his world, a revolutionary assertion that made it possible to program and change society.

Enlightenment

Many philosophers, including Jean-Jacques Rousseau, Voltaire, and Denis Diderot, developed new social ideas during the Enlightenment period that were based on reason and methods of scientific inquiry. These ideas did not draw on ideas of the past from classical thinkers, nor involved "blindly" following religious teachings and authority of the monarch. An important idea was that with new discoveries challenging the status quo way of thinking, scientists were required to find new normativity. This process allowed scientific knowledge and society to progress. French thought during this period focused on moral critique and criticisms of the monarchy.

Modernity arose during the Enlightenment period, with the emergence of the world economy and exchange among diverse societies, bringing sweeping changes and new challenges for society. During the Eighteenth century, many French and Scottish intellectuals and philosophers embraced the idea of progress and ideas of modernity.

Adam Smith addressed the question of whether vast inequalities of wealth represented progress. He explained that the wealthy often demand convenience, employing numerous others to carry out labor to meet their demands. He argued that this allows wealth to be redistributed among inhabitants, and for all to share in progress of society. Smith explained that social forces could regulate the market economy with social objectivity and without need for government intervention. Smith regarded the division of labor as an important factor for economic progress. John Millar suggested that improved status of women was important for progress of society. Millar also advocated for abolition of slavery, suggesting that personal liberty makes people more industrious, ambitious, and productive.

Voltaire's *Lettres Philosophiques* presented new scientific and philosophical ideas developed by Isaac Newton, John Locke, and others, introducing them to the French. Methods used to study scientific phenomenon were extended to study social and moral issues. David Hume used this approach in his *Treatise of Human Nature*na lie

Social Questions

Philosophical questions addressed by social thinkers often centered around modernity, including:

- Can human reason make sense of the social world and shape it for the better?
- Did the development of modern societies, with vast inequalities in wealth among citizens, constitute progress?
- How do particular government interventions and regulations impact natural social processes?
- Should the economy/market be regulated or not?

Other issues relating to modernity that were addressed by social thinkers include social atomization, alienation, loneliness, social disorganization, and secularization.

European Social Thought

Adam Ferguson, Montesquieu, and John Millar, among others, were the first to study society as distinct from political institutions and processes. In the nineteenth century, the scientific method was introduced into study of society, which was a significant advance leading to development of sociology as a discipline.

At the time of the Enlightenment, European societies were still largely rural, with minimal involvement of government in everyday life of citizens. With industrialization and urbanization, societies were significantly transformed, and new ways of thinking about society arose.

In the 19th century, questions involving social order gained importance. The French Revolution freed French society of control by the monarchy, with no effective means of maintaining social order until Napoleon came to power.

Classical Social Theory

The first "modern" social theories (known as classical theories) that begin to resemble the analytic social theory of today developed almost simultaneously with the birth of the science of sociology. Auguste Comte (1798–1857), known as the "father of sociology" and regarded by some as the first philosopher of science, laid the groundwork for positivism – as well as structural functionalism and social evolutionism. In the 19th century three great classical theories of social and historical change emerged: the social evolutionism theory (of which Social Darwinism forms a part), the social cycle theory and the Marxist historical materialism theory.

Another early modern theorist, Herbert Spencer (1820–1903), coined the term "survival of the fittest". Some Post-Modern social theorists like Shepard Humphries, draw heavily upon Spencer's work and argue that many of his observations are timeless (just as relevant in 2008 as 1898). Vilfredo Pareto (1848–1923) and Pitirim A. Sorokin argued that 'history goes in cycles', and presented the social cycle theory to illustrate their point. Ferdinand Tönnies (1855–1936) made *community* and *society* (*Gemeinschaft and Gesellschaft*, 1887) the special topics of the new science of "sociology", both of them based on different modes of will of social actors.

Most of the 19th century pioneers of social theory and sociology, like Saint-Simon, Comte, Marx, John Stuart Mill or Spencer, never held university posts. In this sense they were broadly regarded as philosophers. Emile Durkheim, however, endeavoured to formally established academic sociology, and did so at the University of Bordeaux in 1895, publishing his *Rules of the Sociological Method*. In 1896, he established the journal *L'Année Sociologique*. Durkheim's seminal monograph, *Suicide* (1897), a case study of suicide rates amongst Catholic and Protestant populations, distinguished sociological analysis from psychology or philosophy.

Many of the classical theories had one common factor: they all agreed that the history of humanity is pursuing a certain fixed path. They differed on where that path would lead: social progress, technological progress, decline or even fall, etc. Social cycle theorists were much more skeptical of the Western achievements and technological progress, however, arguing that progress is but an illusion of the ups and downs of the historical cycles. The classical approach has been criticized by many modern sociologists and theorists, among them Karl Popper, Robert Nisbet, Charles Tilly and Immanuel Wallerstein.

Karl Marx rejected Comtean positivism but nevertheless aimed to establish a *science of society* based on historical materialism, becoming recognised as a founding figure of sociology posthumously. At the turn of the 20th century, the first wave of German sociologists, including Max Weber and Georg Simmel, developed sociological antipositivism. The field may be broadly recognised as an amalgam of three modes of social scientific thought in particular; Durkheimian sociological positivism and structural functionalism, Marxist historical materialism and conflict theory, and Weberian antipositivism and verstehen critique.

French Social Thought

- Claude Henri Saint-Simon
- Auguste Comte
- Émile Durkheim

British Social Thought

British social thought, with thinkers such as Herbert Spencer, addressed questions and ideas relating to political economy and social evolution. The political ideals of John Ruskin were a precursor of social economy (*Unto This Last* had a very important impact on Gandhi's philosophy).

German Social Thought

Important German philosophers and social thinkers included Immanuel Kant, Georg Wilhelm Friedrich Hegel, Karl Marx, Max Weber, and Georg Simmel.

Italian Sociology

Important Italian social scientists include Antonio Gramsci, Gaetano Mosca, Umberto Eco, and Vilfredo Pareto.

Chicago School

The Chicago school developed in the 1920s, through the work of Albion Woodbury Small, W. I. Thomas, Ernest W. Burgess, Robert E. Park, Ellsworth Faris, and other sociologists at the University of Chicago. The Chicago school included focus on patterns and arrangement of social phenomenon across time and place, and within context of other social variables. George Herbert Mead, a member of the Philosophy department at the University of Chicago, was also influential.

Theories

Critical Theory

Critical theorists including Hegel rejected the “objective”, scientific approach. They sought to frame theories within ideologies of human freedom.

Marxism

Karl Marx wrote and theorized about the importance of political economy on society, and focused on the “material conditions” of life. His theories strongly centered around capitalism and its affect on class-struggle between the proletariat and bourgeoisie.

Other Perspectives

Other theories include:

- Social constructionist theory
- Rational choice theory
- Structural functionalism – influenced by Spencer and Durkheim
- Social action – influenced by Weber and Pareto
- Conflict theory – influenced by Marx, Simmel
- Symbolic interaction – influenced by George Herbert Mead
- False necessity

Postmodernism

Postmodernism was defined by Jean-François Lyotard as “incredulity towards metanarratives” and contrasted that with modern which he described as “any science that legitimates itself with reference to a metadiscourse... making an explicit appeal to some grand narrative, such as the dialectics of Spirit the hermeneutics of meaning, the emancipation of the rational or working subject, or the creation of wealth.

History

Modern Social Theory

Much of 19th-century classical social theory has been expanded upon to create newer, more contemporary social theories such as multilineal theories of evolution (neoevolutionism, sociobiology, theory of modernization, theory of post-industrial society) and various strains of Neo-Marxism.

In the late 19th and early 20th centuries, social theory became most closely related to academic sociology while other related studies such as anthropology, philosophy, and social work branched out into their own disciplines. Such subjects as "philosophy of history" and other such multi-disciplinary subject matter became part of social theory as taught under sociology.

Attempts to recapture a space for discussion free of disciplines began in earnest in the late 1920s and early 1930s. The Frankfurt Institute for Social Research provides the most successful historical example. The Committee on Social Thought at the University of Chicago followed in the 1940s. In the 1970s, programs in Social and Political Thought were established at Sussex and York. Others followed, with various emphases and structures, such as Social Theory and History (University of California, Davis). Cultural Studies programs, notably that of Birmingham University, extended the concerns of social theory into the domain of culture and thus anthropology. A chair and undergraduate program in social theory was established at the University of Melbourne and a number of universities now specialize in social theory (UC-Santa Cruz is one example). Social theory at present seems to be gaining more acceptance as a classical academic discipline.

Post-modern Social Theory

Scholars most commonly hold *postmodernism* to be a movement of ideas arising from, but also critical of elements of modernism. The wide range of uses of this term resulted in, different elements of modernity are chosen as being continuous. As the different elements of modernity are held to be critiqued. Each of the different uses also is rooted in some argument about the nature of knowledge, known in philosophy as epistemology. Individuals who use the term are arguing that either there is something fundamentally different about the transmission of meaning, or that modernism has fundamental flaws in its system of knowledge.

The argument for the necessity of the term states that economic and technological conditions of our age have given rise to a decentralized, media-dominated society. These ideas are simulacra, and only inter-referential representations and copies of each other, with no real original, stable or objective source for communication and meaning. Globalization, brought on by innovations in communication, manufacturing and transportation. Globalization itself is often cited as one force which has driven the decentralized modern life, creating a culturally pluralistic and interconnected global society, lacking any single dominant center of political power, communication, or intellectual production. The postmodern view is that inter-subjective knowledge, and not objective knowledge is the dominant form of discourse under such conditions, and the ubiquity of copies and dissemination fundamentally alters the relationship between reader and what is read, between observer and the observed, between those who consume and those who produce. Not all people who use the term postmodern or postmodernism see these developments as positive. Users of the term often argue that their ideals have arisen as the result of particular economic and social conditions, including what is described as "late capitalism" and the growth of broadcast media, and that such conditions have pushed society into a new historical period.

The term "postmodernism" was brought into social theory in 1971 by the Arab American Theorist Ihab Hassan in his book: *The Dismemberment of Orpheus: Toward a Postmodern Literature*. In 1979 Jean-François Lyotard wrote a short but influential work *The Postmodern Condition: A report on knowledge*. Jean Baudrillard, Michel Foucault, and Roland Barthes were influential in the 1970s in developing postmodern theory.

Social Theory Today

In the past few decades, largely in response to postmodern critiques, social theory has begun to stress free will, individual choice, subjective reasoning, and the importance of unpredictable events in place of deterministic necessity. Rational Choice Theory and Symbolic Interactionism furnish two examples of more recent developments. False necessity is another. A not uncommon view among contemporary sociologists is that there are no great unifying 'laws of history', but rather smaller, more specific, and more complex laws that govern society.

Philosopher and politician Roberto Mangabeira Unger has more recently attempted to revise classical social theory by exploring how things fit together, rather than to provide an all encompassing single explanation of a universal reality. He begins by recognizing the key insight of classical social theory of society as an artifact, and then by discarding the law-like characteristics forcibly attached to it. Unger argues that classical social theory was born proclaiming that society is made and imagined, and not the expression of an underlying natural order, but at the same time its capacity was checked by the equally prevalent ambition to create law-like explanations of history and social development. The human sciences that developed claimed to identify a small number of possible types of social organization that coexisted or succeeded one another through inescapable developmental tendencies or deep-seated economic organization or psychological constraints. Marxism is the star example.

Calling his efforts "super-theory," Unger has thus sought to develop a comprehensive view of history and society, but to do so without subsuming deep structure analysis under an indivisible and repeatable type of social organization or with recourse to lawlike constraints and tendencies. His most forceful articulation of such a theory is in *False Necessity: anti-necessitarian social theory in the service of radical democracy,* where he employs deep-logic practice to theorize human social activity through anti-necessitarian analysis.

Unger begins by formulating the theory of false necessity, which claims that social worlds are the artifact of human endeavors. There is no pre-set institutional arrangement that societies must adhere to, and there is no necessary historical mold of development that they will follow. Rather, we are free to choose and to create the forms and the paths that our societies will take. However, this does not give license to absolute contingency. Rather, Unger finds that there are groups of institutional arrangements that work together to bring about certain institutional forms—liberal democracy, for example. These forms are the basis of a social structure, which Unger calls formative context. In order to explain how we move from one formative context to another without the conventional social theory constraints of historical necessity (e.g. feudalism to capitalism), and to do so while remaining true to the key insight of individual human empowerment and anti-necessitarian social thought, Unger recognized that there are an infinite number of ways of resisting social and institutional constraints, which can lead to an infinite number of outcomes. This variety of forms of resistance and empowerment make change possible. Unger calls this empowerment negative capability. Unger is clear to add, however, that these outcomes are always reliant on the forms from which they spring. The new world is built upon the existing one.

Theory Construction

Selecting or creating appropriate theory for use in examining an issue is an important skill for any researcher. Important distinctions: a *theoretical orientation* (or paradigm) is a worldview, the lens through which one organizes experience (i.e. thinking of human interaction in terms of power or exchange); a *theory* is an attempt to explain and predict behavior in particular contexts. A theoretical orientation cannot be proven or disproven; a theory can. Having a theoretical orientation that sees the world in terms of power and control, one could create a theory about violent human behavior which includes specific causal statements (e.g. being the victim of physical abuse leads to psychological problems). This could lead to an hypothesis (prediction) about what one expects to see in a particular sample, e.g. "a battered child will grow up to be shy or violent." One can then test my hypothesis by looking to see if it is consistent with data in the real world. One might, for instance, review hospital records to find children who were abused, then track them down and administer a personality test to see if they show signs of being violent or shy. The selection of an appropriate (i.e. useful) theoretical orientation within which to develop a potentially helpful theory is the bedrock of social science.

Lazy User Model

The lazy user model of solution selection (LUM) is a model in information systems proposed by Tétard and Collan that tries to explain how an individual selects a solution to fulfill a need from a set of possible solution alternatives. LUM expects that a solution is selected from a set of available solutions based on the amount of effort the solutions require from the user – the user is supposed to select the solution that carries the least effort. The model is applicable to a number of different types of situations, but it can be said to be closely related to technology acceptance models.

The model draws from earlier works on how least effort affects human behaviour in information seeking and in scaling of language.

Earlier research within the discipline of information systems especially within the topic of technology acceptance and technology adoption is closely related to the lazy user model.

The Model Structure

The model starts from the observation that there is a "user need", i.e. it is expected that there is a "clearly definiable, fully satisfiable want" that the user wants satisfied (it can also be said that the user has a problem that he/she wants solved). So there is a place for a solution, product, or service.

The user need defines the set of possible solutions (products, services etc.) that fulfill the user need. The basic model considers for simplicity needs that are 100% satisfiable and services that 100% satisfy the needs. This means that only the solutions that solve the problem are relevant. This logically means that the need defines the possible satisfying solutions – a set of solutions (many different products/services) that all can fulfill the user need. LUM is not limited to looking at one solution separately.

All of the solutions in the set that fulfill the need have their own characteristics; some are good and suitable for the user, others unsuitable and unacceptable – for example, if the user is in a train and wants to know what the result from a tennis match is right now, he/she may only use the types of solutions to the problem that are available to him/her. The "user state" determines the set of available/suitable solutions for the user and thus limits the (available) set of possible solutions to fulfill the user need. The user state is a very wide concept, it is the user characteristics at the time of the need. The user state includes, e.g., age, wealth, location ... everything that determines the state of the user in relation to the solutions in the set of the possible solutions to fulfill the user need.

The model supposes that after the user need has defined the set of possible solutions that fulfill the user need and the user state has limited the set to the available plausible solutions that fulfill the user need the user will "select" a solution from the set to fulfill the need. Obviously if the set is empty the user does not have a way to fulfill the need. The lazy user model assumes that the user will make the selection from the limited set based on the lowest level of effort. Effort is understood as the combination of monetary cost + time needed + physical/mental effort needed.

Considerations

The lazy user theory has interesting implications when thinking about the effect of learning in technology adoption (for example in the adoption of new information systems).

Chain-linked Model

The chain-linked model or Kline model of innovation was introduced by Stephen J. Kline in 1985, and further described by Kline and Rosenberg in 1986. The chain-linked model is an attempt to describe complexities in the innovation process.

Description

In the chain-linked model, new knowledge is not necessarily the driver for innovation. Instead, the process begins with the identification of an unfilled market need. This drives research and design, then redesign and production, and finally marketing, with complex feedback loops between all the stages. There are also important feedback loops with the organization's and the world's stored base of knowledge, with new basic research conducted or commissioned as necessary, to fill in gaps.

It is often contrasted with the so-called linear model of innovation, in which basic research leads to applied development, then engineering, then manufacturing, and finally marketing and distribution.

Applications

The Kline model was conceived primarily with commercial industrial settings in mind, but has found broad applicability in other settings, for example in military technology development. Variations and extensions of the model have been described by a number of investigators.

Domestication Theory

Domestication theory is an approach in science and technology studies and media studies that describes the processes by which innovations, especially new technology is 'tamed' or appropriated by its users. First, technologies are integrated into everyday life and adapted to daily practices. Secondly, the user and its environment change and adapt accordingly. Thirdly, these adaptations feedback into innovation processes in industry, shaping the next generation of technologies and services. The theory was initially developed to help understand the adoption and use of new media technologies by households (Silverstone et al. 1992), but has since been expanded in the innovation literature as a tool to understand technologies and innovations entering any consuming unit (workplace, country etc. e.g. Lie et al., Habib, Punie, Sørensen) that can be analysed economically, culturally and sociologically. The domestication approach considers both the practical and the symbolic aspects of the adoption and use of technologies, showing how these two elements- the meanings of things, and their materiality, are equally important understanding how technologies become part of everyday life. It is a foremost a social theory as it highlights the negotiations, challenges to power and control, rule-making and breaking that accompany the introduction of technologies into any social setting.

The Domestication approach has roots in cultural studies of media use, but is informed by Science and technology studies, gender studies of household technology, sociology of everyday life, consumption studies and innovation studies, and has been most widely used in studying the mass adoption of computers, internet and mobile phones.

As a strand of the Social shaping of technology approach to understanding how technology is created, Domestication theory highlights the role of users in innovation - the work done by individuals and communities in order to make a technology from the outside do practical work, and make sense within that community. This strand of work links to the role of end users, lead users etc. in long term innovation processes (Williams et al. 2004).

Domestication studies are generally done using qualitative methods, such as long interviews and ethnography to explore the emerging meanings of technologies, and the changing routines, and conflicts that would not normally be accessible to quantitative methods.

The Domestication approach uses a number of different concepts to distinguish various aspects of the process. For example: Appropriation is the process of bringing a technology into a household, or another local cultural context; Conversion is the remaking of the meanings, or values and norms associated with the technology and the transfer of these back to the 'outside' world.

The principal criticism of the domestication approach is its reliance of detailed case studies, and its rather descriptive approach which is difficult to turn into prescriptive lessons of the type required by business and policy makers. However, this rich-descriptive approach is also its strength: it enables processes and the complex interplay of artefacts and cultural values to be explored in much more depth than individualistic, quantitative methods.

The Domestication approach, describing the integration of technologies into social relationships and structures using evidence obtained using qualitative methods, stands in sharp contrast to individualistic and quantitative approaches (such as Technology acceptance model) of North-American marketing and IS research, that draw on primarily psychological models.

References

- Noel, Hayden (2009). Consumer behaviour. Lausanne, Switzerland La Vergne, TN: AVA Academia Distributed in the USA by Ingram Publisher Services. ISBN 9782940439249.
- Loudon, David L.; Bitta, Albert J. Della (1993). Consumer behavior: concepts and applications. McGraw-Hill Series in Marketing (4th ed.). New York: McGraw-Hill. ISBN 9780070387584.
- Rogers, Everett M. (1983). Diffusion of innovations (3rd ed.). New York: Free Press of Glencoe. ISBN 9780029266502.
- Antonelli, C. (2003). The Economics of Innovation, New Technologies, and Structural Change. London: Routledge. ISBN 0415406439.
- Boserup, Ester (2005-01-01). The Conditions of Agricultural Growth: The Economics of Agrarian Change Under Population Pressure. Transaction Publishers. ISBN 9780202307930.
- Macionis, John J.; Plummer, Ken (2005). Sociology. A Global Introduction (3rd ed.). Harlow: Pearson Education. p. 12. ISBN 0-13-128746-X.
- Unger, Roberto (2004). False Necessity: Anti-Necessitarian Social Theory in the Service of Radical Democracy, Revised Edition. London: Verso. pp. 35–36, 164, 169, 278–80, 299–301. ISBN 978-1-85984-331-4.
- Barsky, Robert (2004). "OIL AND THE MACROECONOMY SINCE THE 1970s" (PDF). Cambridge. Retrieved 2015-11-01.
- T. Carson, Richard. "The Environmental Kuznets Curve: Seeking Empirical Regularity and Theoretical Structure" (PDF). Retrieved 2015-11-01.
- Hayami, Yujiro (1971). "Induced Innovation in Agricultural development" (PDF). conservancy. Retrieved 2015-11-01.

5

Methods and Techniques of Innovation Management

This text discusses the methods and techniques of innovation management in a critical manner providing key analysis to the subject matter. Some of the methods explained in section are innovation leadership, creativity techniques, systematic inventive thinking, brainstorming, nominal group technique, social entrepreneurship etc. The chapter serves as a source to understand the major categories relate to innovative management.

Innovation Leadership

Innovation leadership is a philosophy and technique that combines different leadership styles to influence employees to produce creative ideas, products, and services. The key role in the practice of innovation leadership is the innovation leader. Dr. David Gliddon (2006) developed the competency model of innovation leaders and established the concept of innovation leadership at Penn State University.

As an approach to organization development, innovation leadership can support achievement of the mission or the vision of an organization or group. With new technologies and processes, it is necessary for organizations to think innovatively to ensure continued success and stay competitive. to adapt to new changes, "The need for innovation in organizations has resulted in a new focus on the role of leaders in shaping the nature and success of creative efforts." Without innovation leadership, organizations are likely to struggle. This new call for innovation represents the shift from the 20th century, traditional view of organizational practices, which discouraged employee innovative behaviors, to the 21st-century view of valuing innovative thinking as a "potentially powerful influence on organizational performance."

Overview

To have a clear understanding of what innovation leadership involves, one must first understand the concept of innovation. Although there is some controversy over how it can be defined, through general consensus in the literature, it can be described as novel ideas of viable products that are put into operation. It includes three different stages, which are all dynamic and iterative (constant):

1. Idea Generation

2. Evaluation
3. Implementation

The two types of innovation include exploratory innovation, which involves generating brand new ideas, and value-added innovation, which involves modifying and improving ideas that already exist. Ideas generated must be useful to be considered innovative. Innovation should also not be confused with creativity, which is merely the generation of a novel idea that may not necessarily be put into operation—though these words are sometimes used interchangeably in research literature when speaking about innovation leadership. Innovation leadership is a complex concept, as there is no single explanation or formula for a leader to follow to increase innovation. As a result, innovation leadership encompasses a variety of different activities, actions, and behaviors that interact to produce an innovative outcome.

Value-added Innovation

Exploratory and value-added innovation require different leadership styles and behaviors to succeed. Value-added innovation (PwC, 2010) involves refining and revising an existing product or service and typically requires minimal risk taking (compared to exploratory innovation, which often involves taking a large risk); in this case, it is most appropriate for a leader for innovation to adopt a transactional form of leadership. This is because a transactional leadership style does not use open leadership behaviors such as encouraging employees to experiment and take risks, but rather uses closed leadership behaviors that do not condone or reward risk-taking. Companies whose innovation leaders use transactional leadership for value-added innovation purposes include Toyota Motor Co., General Motors Corp., and Ford Motor Co.; examples of these companies' value-added innovations such as making improvements on existing cars by making them faster, more comfortable, and getting better gas mileage.

Occasionally a value-added innovation may require a completely new way of thinking and possibly taking new risks. An example of this scenario can be illustrated through Aspirin; this was an existing product, traditionally used as an analgesic to alleviate aches and pains, but has been introduced into a new and different market by extending its uses to help prevent heart attack and reduce blot clot formation. In this example, the usage of an existing product was re-worked and introduced into a new market. While an existing product is being changed and/or improved upon, characterizing it as a value-added innovation, outside-the-box thinking, research, and risk-taking are now required since it is being introduced into a new market. In this case, a transformational leadership style is a more appropriate style to use.

The innovation leader must gauge if (and how much) risk and radical thinking are involved in the value-added innovation to determine which leadership style to use in a situation. The leader must be flexible—able to switch leadership behaviors when necessary.

Exploratory Innovation

Exploratory innovation refers to the generation of novel ideas, strategies, and solutions through the use of strictly open behaviors exhibited most often by transformational leaders. The foundation of exploratory innovation is characterized by search, discovery, experimentation, and risk taking. It is the organization's focus on generating new ideas, products and strategies; in contrast to exploitative innovation, which focuses on building and extending already existing ideas. Some studies have shown that explorative and exploitative innovation require different structures, strategies, processes, capabilities, and cultures. Exploratory innovation requires flexibility, opportunism, adaptability, and for leaders to provide intellectual stimulation to their subordinates. In this approach to innovation, the leadership style that is primarily used is transformational. The behaviors exhibited are believed to achieve the desired creative outcome from employees through the application of individualized consideration, charisma, and inspirational motivation.

For example, in one study of the innovation practices at AXA Insurance in Ireland, the CEO John O'Neil engaged in transformational leadership behaviors and introduced the "MadHouse" program that combined workers from different departments and levels of the organization to work together in a creative way. The result of this experiment after six months was 150 new business ideas for products and services. Explorative and Value-added innovation are often referenced together, but surprisingly little research shows an interaction between the two. However, there is an understanding that in some circumstances a 'balance' must be attained to achieve superior employee performance. For example, not all novel ideas are implemented, and may be resurrected later. The organization may need to switch gears and adopt exploitative strategies to revise and refine the idea to match present needs.

Foundations of Innovation Leadership

Innovation leadership has roots in path-goal theory and leader-member exchange theory. Certain elements within an organization are also needed for innovation leadership to succeeded. Wolfe (1994), as cited by Sarros, Cooper, & Santora, (2008) has pointed out that one antecedent factor for innovation is organizational culture. Likewise, Isaksen, Laver, Ekvail & Britz (2001) concur that innovative endeavors fail without a supportive climate. This antecedent of a supportive organizational culture/climate encompasses encouragement of creativity, autonomy, resources, and pressures. Additional foundational elements for innovation leadership include creative work, a creative workforce, and certain leader attributes.

Roots in Path-Goal Theory

The basis of path-goal theory uses a similar view of leadership, in that it advocates different types of leadership (e.g., participative, supportive) behaviors, much like innovation leadership does. However, it is contingent on employee and environmental factor to be effective. The idea of a single leader using different leadership behaviors originated in path-goal theory, and has been associated with the framework underlying innovation leadership, which also allows the creation of a work environment conducive to innovative thinking—which is the cognitive process of generating novel and useful ideas.

Creating this type of work environment through innovation leadership involves open leadership behaviors that resemble some leader behaviors proposed by Path-goal theory—for example, upward influence and supportive/considerate behaviors. In innovation leadership, these behaviors encourage the creative team to generate as many novel ideas as possible and lead to evaluation and implementation of these ideas.

Roots in Leader-Member Exchange Theory

Leader-member exchange theory (LMX theory) is another one of the building blocks of innovation leadership. It follows the same idea as Path-goal theory and innovation leadership, that multiple leadership styles are necessary in managing multiple subordinates but takes it a step further. LMX involves adopting a unique leadership style for each employee. Past studies indicate that LMX theory has been shown to have an effect on innovation. Studies have also shown that leader-member exchange relationships can predict significant organizational and attitudinal variables including higher job satisfaction and higher job performance.

Basu and Green (1997) found that innovative behavior is related to the quality of the leader-member exchange where high quality exchanges include contributions from both the leader and the follower. However, in a study by Jean Lee (2008), only the loyalty aspect of LMX (LMXL) was shown to be related to innovativeness. Leadership styles, transformational (positively related) and transactional (negatively related), were found to have an effect on innovativeness.

Innovative Organizational Culture/Climate

Some studies have shown evidence of organizational culture as the mediator of the relationship between transformational leadership and organizational innovation and performance. In other words, for transformational leadership to affect organizational innovation, an organization must have a strong innovative culture in addition to a leader with a transformational leadership style.

Organizational culture refers to an organization's deep structure, normative beliefs, and shared behavioral expectations. This culture is fairly constant and can influence interorganizational relations. Climate refers to the way that individuals perceive the extent to which the organizational culture impacts them. The two essentially are interrelated. One proposed model for assessing a creative environment in organizations includes the following dimensions:

Encouragement of Creativity

Encouragement of creativity is the most frequently mentioned dimension in the literature. It operates at three major levels, each level containing multiple aspects.

Conceptual Model Underlying Assessment of Perceptions of the Work Environment for Creativity

Organizational Encouragement

The first level is Organizational Encouragement. This involves encouragement of risk-taking and idea generation from all levels of management, fair and supportive evaluation of new ideas, recognition and reward of creativity, and collaborative idea flow across an organization. Each of these are equally important aspects of organizational encouragement but the third aspect, recognition and reward of creativity, may have adverse effects if the sole purpose for engaging in an activity is to gain reward.

Supervisory Encouragement

The second level, Supervisory Encouragement, highlights the roles of supervisors and project managers in goal clarity, open interaction between supervisors and subordinates, and supervisory support of a team's work and ideas. This level of encouragement points to the concepts of transformational leadership and LMX that emphasise the importance of the interactions of supervisors and subordinates in innovative performance.

Work Group Encouragement

The third level of encouragement is Work Group Encouragement. Diversity in team members' backgrounds and openness to ideas affects creativity because individuals are exposed to a variety of novel and unusual ideas and such exposure had been demonstrated to have a positive effect on creative thinking

Autonomy

Autonomy is believed to foster creativity since studies have revealed that individuals produce more creative work and experience increased intrinsic motivation when they have a sense of control and ownership over their work and ideas and they perceive themselves as having a choice in how their goals are accomplished, whether those goals are given to them by their supervisor or chosen by themselves.

Resources

Resources have been suggested to be directly related to creativity in organizations. Individuals' perceptions of the availability of resources may lead to increased beliefs of the likelihood that the ideas they generate have a possibility of reaching the implementation stage.

Pressures

The little evidence that exists on the dimension of pressure suggests somewhat paradoxical influences. Some degree of pressure could have a positive effect if the pressure originates from the challenging and intellectual nature of the task itself, increasing intrinsic motivation. However, if the pressure experienced is perceived extreme it could counteract creativity. Amabile et al. (1996) identify two forms of pressure: excessive workload pressure and challenge. They suggest that the first should have a negative influence on creativity while the second should have a positive one.

Organizational Impediments to Creativity

Although there has been little research on the work of environment factors that undermine creativity, some research suggests that these impediments include internal strife, conservatism, and rigid, formal management structures within organizations. This dimension is seen as working against autonomy and tends to have an inverse effect as individuals may perceive a more controlling environment.

Creative Work

Creative work can occur at any job but more specifically it can occur in jobs where there are complicated, ill-defined problems requiring innovative solutions. The fact that the creative problems are ill-defined makes the creative work uncertain, and it may involve risky efforts. It is also resource intensive, demanding, and time consuming work requiring high levels of motivation and often requires collaboration. This type of work must also involve both novel idea generation and novel idea implementation and requires expertise on the part of the workforce.

Creative Workforce

A creative workforce is needed for innovation leadership to be successful. Creative people have expertise on the subject requiring innovation and tend to use work as a source of identity. Because of this, they are powerfully intrinsically motivated by professional achievement opportunities and recognition. Creative workers are also commonly characterized as highly valuing their autonomy; additional dispositional attributes include openness, flexibility, cognitive complexity, self-confidence, dominance, and introversion. The patterns of characteristics creative workers exhibit typically allow them to confidently explore alternative ideas under ambiguous conditions.

Leader Attributes/Characteristics

Successful innovation leadership requires a leader with certain characteristics. These include expertise in the domain, creativity, ability to carry out transformational leadership behaviors, planning and sense-making, and social skills. Innovative leaders can be recruited and hired through professional networks and referrals or alternatively found through succession planning, which involves identifying innovative leaders who are already working within the organization.

Types of Innovation Leadership Styles

In addition to these foundations, various styles play an important role in innovation leadership, each of which are used at different stages of the innovation process or for different types of innovation (value-added vs. exploratory). Frequently associated leadership styles include transformational leadership, transactional leadership, and Ambidextrous leadership. The type of leadership most strongly associated with innovation is transformational leadership.

Key Activities of Innovation Leadership

Idea Generation

As mentioned above, different leadership styles and behaviors may be more appropriate at different stages of the innovation process. Current research supports the notion that in the idea generation process, innovation leadership requires a leader to use a more transformational style of leadership. During this stage, a leader needs to promote a safe environment for employees/team members to voice novel ideas and original thinking as well as provide workers with the resources to do so effectively. Research has also found that leaders who engage in unconventional behaviors, associated with transformational leadership, were seen as stronger role models and, as a result, increase creative performance in their subordinates. For example, the founders of Google have been known to wear capes and jump-shoes around the office, thus inspiring more outside-the-box thinking in their employees. These open leadership behaviors convey that unorthodox and unconventional ideas and behaviors are not only accepted but also encouraged.

Idea Evaluation and Implementation

In addition to providing a climate for idea generation, innovation leadership also requires leaders to ensure that the process of idea generation does not overshadow the evaluation and implementation processes. During these phases of leadership, leaders must support some ideas while discarding other ideas and put the supported ideas into production. The role of the leader must shift away from a transformational style to a more transactional style of leadership, which involves being more direct and critical toward the ideas generated. A leader now needs to ensure that constructive discussions of innovative ideas are taking place among their subordinates. This serves to evaluate the usefulness of each idea, eliminate those that do not appear viable to the organization or goal, and push the ones that do appear viable into the production phase. The leader must adopt what are known as closed leadership behaviors to achieve this. Instead of stimulating idea generation, the leader must shift focus from generating new ideas toward fine-tuning existing ideas to achieve progress toward the goal, and ultimately implement the idea. This challenge of balancing differing leadership styles when appropriate is called the *generator evaluator paradox*. It is important to consider the role of ambidextrous leadership, since a leader must be able to switch between leadership roles and styles when necessary to successfully lead for innovation. Paradoxes of innovation leadership are discussed below.

Innovation Leadership and Influence

Depending on the type of leadership style that is adopted by the innovation leader, the leader may have either a direct or indirect influence on your employees.

Direct Influences

Direct forms of influence in leading innovation include:

- providing creative input and idea suggestion to employees
- providing employees with clear and concrete goals
- allocating organizational resources (i.e. research and development spending; manpower) for implementing ideas

Indirect Influences

Indirect influences get the same results without providing explicit guidance to employees. These types of influences include:

- establishing a supportive climate for creativity within the organization
- acting as a role model for innovative thinking
- providing employees with rewards and recognition for innovative thinking
- hiring and team composition (i.e. putting together teams with specific skill sets needed for innovative thinking, or hiring employees with creative personalities without planning what they work on).

Proposed Model for Innovation Leadership

A proposed model for innovation leadership has been a multilevel process model of innovation, which uses the direct and indirect leadership on the processes of innovation mentioned in the above section to promote the innovation process. In the model, indirect leadership influences affect the individual creativity (generation phase) and team creativity (evaluation phase) process. Direct leadership influences affect the team creativity (evaluation phase) process and the organizational innovation process (implementation phase). The individual creativity (generation phase) box in the model represents the process of the individual generating the initial idea or ideas and proposing them to their team. The team creativity (evaluation phase) box represents the process of the team taking that idea, making alterations and fine-tuning it to the point of making prototypes, formalized sketches, or simulations. The organizational innovation (implementation) box represents taking those prototypes, sketches, or simulations and testing, evaluating, and possibly mass-producing them.

Figure 1. Model of direct and indirect leadership influences on the process of innovation

Two very important key features of this model should be mentioned:

1. The three stages of innovation (idea generation, evaluation, and implementation) are not independent of one another.
2. The stages in the model should not be viewed in a "lock-step fashion," meaning that there are both backward and forward influences and activities affecting each of the three stages. For example, ideas are generated, discussed, and tested only to feed information back into the system, starting the process from the beginning again. The forward and backward arrows between individual creativity and team creativity, the forward and backward arrows between team creativity and organizational innovation, as well as the arrow from organizational innovation to individual creativity visually represent this key feature.

Paradoxes of Innovation Leadership

Innovation leadership is complex, as can be seen from the Hunter & Cushenbery (2011) model, and often paradoxes emerge that require leaders to strike a delicate balance between two conflicting roles (e.g. encouraging innovative ideas vs. limiting innovative ideas to include only those that are most viable and useful to the organization). A balance must be struck, not only within the leader and their behaviors, but between conflicting interests of involved parties as well. These include conflicting interests between the leader and the employees/teams, between leaders and situational/contextual factors, and between the employees/teams and the organization. Critical potential paradoxes that are often faced by leaders of innovation have been provided by Hunter, Thoroughgood, Meyer, & Ligon (2011).

Internal/Localized Paradox

Internal/Localized paradoxes entail conflicting roles experienced within the leader.

Dual Expertise Paradox

The Dual Expertise Paradox postulates that a leader must have or acquire domain expertise while at the same time obtaining the necessary leadership skills to manage his/her employees and resources.

Generation Evaluation Paradox

The Generation Evaluation Paradox stipulates that a leader must encourage a supportive climate for the generation of new ideas and thinking outside-the-box while evaluating these ideas and realizing that not all creative ideas are useful and many may even fail (while not being too critical and negative of those ideas).

Team-level Paradox

Team-level paradoxes entail conflicting interests between the leader and the employees/teams

Creative Personality Cohesion Paradox

Creative Personality Cohesion Paradox is based on the research finding that creative workers generally highly value autonomy and, as a result, often prefer to work alone. This paradox illustrates the difficulty leaders have in providing their employees with the autonomy they must be creative, while fostering team cohesion (or closeness) to facilitate idea sharing. A leader must also be careful not to encourage too much cohesion, as it may discourage group members from disagreeing (even constructively disagreeing) with fellow group members in an effort not to offend them or "rock the boat."

Vision Autonomy Paradox

The Vision Autonomy Paradox highlights the dilemma a leader faces between providing structure and guidance to a team with respect to the vision of the goal, while at the same time stepping back and providing the team with enough autonomy, especially considering the fact that creative workers highly value autonomy. When leading for innovation, providing an overabundance of structure may result in a backlash from employees who feel their autonomy is being taken away from them.

Restriction Freedom Paradox

The Restriction Freedom Paradox underscores that innovation leaders need to allow employees enough time to develop creative endeavors and provide the resources to do so. At the same time the leader must take care to provide enough pressure that they are still motivated to complete the task and not provide too many resources that it has a "deadening effect" on creativity.

Situational Paradox

Situational Paradoxes entail conflicting interests between leaders and the situations they face.

Intrinsic Extrinsic Paradox

The Intrinsic Extrinsic Paradox holds that instead of providing more readily available extrinsic motivation tools such as bonuses and salary increases, leaders must provide intrinsic motivation, which generally comes from within the employee, to their employees. This paradox is based on findings that intrinsic motivation is a key factor in facilitating creativity and extrinsic motivators may either hinder creativity or have an unclear relationship with creativity.

Local Long-Term Paradox

The Local Long-Term Paradox posits that leaders of innovation must maintain their innovative edge by keeping an eye out for and capitalizing on potential opportunities, even at the risk of placing those ideas above or even eliminating ideas that he or she had previously inspired in their teams. The leader must also be capable of developing teams that are flexible enough to be passionate about ideas that may have replaced their own idea that was facilitated, inspired, and supported by their leader. This is where the paradox is most clearly visible.

Competition Collaboration Paradox

The competition collaboration paradox involves a leader developing open external relationships with other organizations to discover potential innovation opportunities, while ensuring the organization's emerging ideas are protected in a competitive environment.

Feedback Rigidity Paradox

The feedback rigidity paradox involves leaders seeking out and using customer and client advice and feedback towards innovative endeavors to a certain extent, while maintaining control of the vision and not letting the feedback dictate to them—as clients and customers often criticize innovations early on.

Failure Success Paradox

The Failure Success Paradox is the idea that innovation leaders must ensure a safe organizational culture that is willing to embrace risk and failure, while at the same time making sure that the organization is also producing successful products and services despite embracing risk and errors.

Additional Paradoxes

Additional paradoxes identified by Hunter et al. (2011) that do not directly involve the leader but are worth mentioning are the paradoxes that occur between teams and the organization. These include the Insularity Cohesion paradox, the Champion Evaluator paradox, and the Creativity Cost paradox.

Outcomes

Outcomes of innovation leadership include inspiring employees to the create and implement novel ideas for products, services, and technologies. In addition, these novel ideas can also be used to solve problems within an organization. What this illustrates is that innovation spurred by innovation leadership can be translated across various industries and can be used for a multitude of purposes. Ultimately, inspiring and initiating organizational innovation through innovation leadership can serve to advance the organization it to the next level.

Real-World Examples of Innovation Leadership

Companies that use innovative leadership include 3M, which lets employees work on a project of their choosing for 15% of their time.

Similarly, Google allows employees one day a week to work on their own project.

Zappos employees are allowed to "radically" decorate their cubicle and are encouraged to laugh and have fun at work impromptu in-office parades.

Creativity Techniques

Creativity techniques are methods that encourage creative actions, whether in the arts or sciences. They focus on a variety of aspects of creativity, including techniques for idea generation and divergent thinking, methods of re-framing problems, changes in the affective environment and so on. They can be used as part of problem solving, artistic expression, or therapy.

Some techniques require groups of two or more people while other techniques can be accomplished alone. These methods include word games, written exercises and different types of improvisation, or algorithms for approaching problems. Aleatory techniques exploiting randomness are also common.

Aleatory Techniques

Aleatoricism is the incorporation of chance (random elements) into the process of creation, especially the creation of art or media. Aleatoricism is commonly found in music, art, and literature, particularly in poetry. In film, Andy Voda made a movie in 1979 called "Chance Chants", which he produced by a flip of a coin or roll of a dice. In music, John Cage, an avant-garde musician, composed music by superimposing star maps on blank sheet music, by rolling dice and preparing open ended scores that depended on the spontaneous decisions of the performers. (1) Other ways of practicing randomness include coin tossing, picking something out of a hat, or selecting random words from a dictionary.

In short, aleatoricism is a way to introduce new thoughts or ideas into a creative process.

Improvisation

Improvisation is a creative process which can be spoken, written, or composed without prior preparation. Improvisation, also called extemporization, can lead to the discovery of new ways to act, new patterns of thought and practices, or new structures. Improvisation is used in the creation of music, theater, and other various forms. Many artists also use improvisational techniques to help their creative flow.

The following are two significant methods:

- Improvisational theater is a form of theater in which actors use improvisational acting techniques to perform spontaneously. Many improvisational ("improv") techniques are taught in standard drama classes. The basic skills of listening, clarity, confidence, and performing instinctively and spontaneously are considered important skills for actors to develop.
- Free improvisation is real-time composition. Musicians of all kinds improvise ("improv") music; such improvised music is not limited to a particular genre. Two contemporary musicians that use free improvisation are Anthony Braxton and Cecil Taylor. Through free improvisation, musicians can develop increased spontaneity and fluency.

Each type of improvisation improves the thinking and acting skills of the actor, and this is done by using no practice. A similar set of techniques is called *alienation* since one of its many techniques uses actors that haven't rehearsed or even read the play. Improvisation is an acting technique during which actors make up a storyline, start and end on the spot, and try their best to keep in character.

Problem Solving

In problem-solving contexts, the random-word creativity technique is perhaps the simplest method. A person confronted with a problem is presented with a randomly generated word, in the hopes of a solution arising from any associations between the word and the problem. A random image, sound, or article can be used instead of a random word as a kind of creativity goad or provocation.

There are many tools and methodologies to support creativity:

- TRIZ (theory which are derived from tools such as ARIZ or TRIZ contradiction matrix)
- Creative Problem Solving Process (CPS) (complex strategy, also known as Osborn-Parnes-process)
- Lateral thinking process, of Edward de Bono
- Six Thinking Hats, of Edward de Bono

- Method Herrmann - right brain / left brain
- Brainstorming and Brainwriting
- Think outside the box
- Business war games, for the resolution of competitive problems
- SWOT analysis
- The method USIT of convergent creativity
- Thought experiment
- Five Ws
- Coaching

Systematic Inventive Thinking

Systematic Inventive Thinking (SIT) is a thinking method developed in Israel in the mid-1990s. Derived from Genrich Altshuller's TRIZ engineering discipline, SIT is a practical approach to creativity, innovation and problem solving, which has become a well known methodology for Innovation. At the heart of SIT's method is one core idea adopted from Genrich Altshuller's TRIZ which is also known as Theory of Inventive Problem Solving (TIPS): that inventive solutions share common patterns. Focusing not on what makes inventive solutions different - but on what they share in common - is core to SIT's approach.

History of the SIT Methodology

Overview - The Creativity Debate

SIT deals with two main areas of creativity: Ideation of new ideas, and problem solving. In the 1970s, researchers from the field of Cognitive Psychology established a quantitative yardstick for measuring creativity: the creative person was defined as someone with a large flow of ideas. A high rate of ideas per unit of time was considered an indication of creativity. This approach led to a series of methods for developing creativity based on the assumption that a quantitative increase of ideas will necessarily bring about a qualitative improvement. Such widely known methods as brainstorming, Synectics, random stimulation and lateral thinking (identified with Edward de Bono) can be traced to this approach. More recent studies reveal the appearance of a different approach. These studies show that the main difficulty faced by problem solvers is not generating a large quantity of ideas but coming up with original ones. The former parallel drawn between quantity and quality appears to no longer hold true. It was discovered that a large flow of ideas does not necessarily lead to the creation of original ones and further, that the very occupation with ordinary ideas may actually hamper creativity and innovative thought. These discoveries have prompted a new approach which holds that original and interesting results stem from organized thinking and structured processes rather than the random generation of ideas. One of the characteristics of organized thinking is a state of "low stimuli", unencumbered by a large quantity of ideas. In this approach, originality replaces quantity as a dominant criterion. This organized or structured approach to idea generation is the starting point for Systematic Inventive Thinking.

The TRIZ Heritage

SIT is a descendant of the work of Genrich Altshuller, a Russian engineer who analyzed over 200,000 patents to identify the 40 common inventive principles of his unique formula, named TRIZ. Altshuller's main discovery was that creative solutions incorporate an elimination of a conflict in the problem state. A conflict is a state where one parameter must be changed, in order to get some benefit, but changing that parameter causes a deterioration of another important parameter. Routine engineering design deals with this situation by searching for the "best fit" compromise, a trade off that maximized the utility and minimizes the negative impact of a specific configuration of the variance of the available input parameters.

Altshuller found that the engineering conflicts can be indexed according to the type of parameters involved (39 common engineering parameters were initially defined).

Examining numerous inventions made it possible to assign each conflict with a set of possible hints or strategies on how to approach the solution to the problem. Three types of hints are used: principles, standards and physical effects. There are 40 principles; each help with defining high level strategies for solving the problem. There are 70 standards, which are more elaborated ideas based on collective past solutions. There is a knowledge base of approximately 400 physical effects ranging across physical, chemical and geometrical aspects, indexed according to the functions that each effect can carry out.

During the 1970s, one of Altshuller's students, Ginadi Filkovsky, immigrated to Israel and joined the Open University in Tel Aviv. He began teaching TRIZ and adapting it to the needs of both Israeli and international hi-tech companies. A number of key academics were involved in this research.

Two Ph.D. students, Jacob Goldenberg and Roni Horowitz, joined Filkovsky, focusing their research on developing and simplifying the methodology. Their work formed the basis of the SIT method as it exists today. Both TRIZ and SIT share a basic assumption - that one can study existing creative ideas in a field, identify common logical patterns in these ideas, translate the patterns into a set of Thinking Tools, and then apply these Thinking Tools to generate new creative ideas. In spite of the commonalties, SIT strongly differs from TRIZ in several important respects, having to do mainly with its practical application.

From TRIZ to SIT

The move from TRIZ to SIT was motivated by the desire to create a method that is easier to learn and retain (achieved through a smaller number of rules and tools), more universal in application (achieved through elimination of engineering specific tools) and tighter in keeping the problem solver within a real inventive framework (the Closed World principle). TRIZ also favors using existing resources for solving a problem. But in contrast to SIT, this principle is scattered around the method. It can be found in the principle of Ideal Final Result ("the best system is when there is no system" - Altshuller). The difference between TRIZ and SIT in this respect is that in SIT the Closed World condition is the most important principle. This is particularly applicable when the template approach is applied to problem solving. The first step in using SIT for Problem Solving is to define the problem world. Once defined, the problem solver knows that all the building blocks for the solution are right there in front of him and that the solution simply requires the reorganization of the existing objects. This adds great focus and power to the method. It also turns every real problem into an amusing puzzle.

Five Thinking Tools

1. Subtraction

Remove an essential component from a product and find uses for the newly envisioned arrangement of the existing components. This abstracted arrangement is known as a 'virtual product'.

2. Multiplication

Add to a product a component of the same type as an existing component. The added component should be changed in some way. The 2 keywords for this tool are: 1) more and 2) different. These represent the two stages for applying the tool: 1) add more copies of something that exists in the product and 2) change those copies according to some parameter.

3. Division

Divide the product and/or its components and rearrange them to form a new product. Using this tool forces consideration of different structures, either on the level of the product/service as a whole, or on the level of an individual component. Dividing a product to many pieces gives the freedom to reconstruct it in many new ways – it increases our Degrees of Freedom for working with the situation.

4. Task Unification

Assign a new and additional task to an existing resource. Less affluent cultures are more likely to adopt the Task Unification mindset. For example, the Bedouins use camels for a number of different tasks: transportation, currency, milk, skin for tents, shade, protection from the wind, burning feces for fuel. More affluent societies tend to jettison resources.

5. Attribute Dependency

Creating and dissolving dependencies between variables of a product. Attribute Dependency works with variables rather than components. Variables are easy to identify as those characteristics that can change within a product or component (e.g. color, size, material).

Principles

The Closed World - Thinking Inside the Box

The Closed World condition is crucial to SIT's methodology. The first step in using SIT is to define the problem world. Once defined, the problem solver knows that all the building blocks for the solution are right there in front of him and that the solution simply requires the reorganization of the existing objects. This adds great focus and power to the method. It also can turn every real problem into an amusing puzzle. The Closed World condition deals with the resemblance between the problem world and the solution world. The condition stipulates that in the development of a new product - or when addressing a problem - one must utilize only elements already existing in the product/problem, or in the immediate environment. This condition forces us to rely on resources already at our disposal, rather than "importing" new external resources for the solution. The Closed World condition often provokes resistance as it runs counter to some of the most common intuitions about creative thinking, especially the ubiquitous notion of "thinking out of the box". The essential claim of "thinking out of the box" is that in order to produce ideas that are new and different, you need to somehow move beyond normal thinking patterns, to a universe located outside the metaphorical box. The Closed World condition, by contrast, forces the thinker to find a creative solution by heavily limiting the space of possibilities. Since the scope of possibilities is artificially limited there is no choice but to reconsider the relations between elements found within the problem and pay closer attention to them: their arrangement in space and time; their assigned functions and their necessity. Thus, the Closed World condition sets us on a collision course with our fixedness, allowing us to arrive at solutions which are both innovative (different from the usual) and simple (since based on existing and known elements).

Qualitative Change

The Qualitative Change principle dictates that solutions can be found where the main problem element or variable in the existing situation is either totally eliminated or even reversed. In other words, a problematic element in a situation is neutralized, so it no longer presents an obstacle. It can also be the case that the problematic element becomes a key positive factor; the situation is 'reversed', and the disadvantage transforms into an advantage.

Function Follows Form

A term coined by Ronald Finke, Function Follows Form is often considered a "backwards" process in that the starting point for thinking of new ideas is the existing resource base rather than specific needs that have been identified in the market. These needs, however, are never ignored – they are simply introduced at a later stage. The process begins with an existing product (or service), continues with manipulating it systematically to create what SIT calls a 'virtual product' (this is the Form) and only then examines whether it satisfies existing or potential customer needs (Function). The Function Follows Form principle is applied as an overarching framework to focus the application SIT's Thinking Tools.

Function Follows Form

Function follows form is a principle associated with classic architecture, and it is also being used in government buildings. The concept of this principle means design created according to a shape, but do not consider its function at initial stage of architectural design.

Path of Most Resistance

In nature, water cascading down a mountain will always follow the path of least resistance – the easiest route. In thinking, too, our minds tend to take the path of least resistance – those avenues most familiar to us. So doing, it is difficult to arrive at ideas new to us or to our competitors. SIT encourages an approach to the counter-intuitive path – the path of most resistance.

Cognitive Fixedness

Cognitive Fixedness is a state of mind in which an object or situation are perceived in one specific way, to the exclusion of any alternative. There are several types of fixedness, among them:

Functional Fixedness

A term coined by the social scientist Karl Dunker. Functional fixedness is the tendency to ascribe specific functions to respective objects. Dunker sees Functional Fixedness as a "Mental block against using an object in a new way that is required to solve a problem.", as described in his cognitive performance test, known as the Candle problem.

Structural Fixedness

This is the tendency to view objects as a whole, as a gestalt, which often makes it difficult to imagine how the product could be reorganized to look differently. For example, why do TV controls always have to be on the lower part of the TV set? Would it not be easier if they were on the upper part? When television sets were first introduced, the controls were potentiometers made out of wax. The heat that was emitted from the Cathode Ray Tubes, dispersing upwards would melt the controls had they been located on the upper part. Therefore, they were placed on the lower part. But since then, new generations of TV sets are available, and still one would find the controls on the lower part. This is Structural Fixedness.

Near Far Sweet

Most ideas for new products are either uninspired or impractical. Finding the "sweet spot" requires a balance that leads to both ingenious and viable ideas. This notion was expressed in a Harvard Business Review Article entitles "Finding Your Innovation Sweet Spot". When we innovate we push our thinking outward, trying to create something new and different from what we have now. Yet we do not want to wander too far off. A "great" idea must be executable and palatable. The NFS principle assures that our ideation will generate ideas that are sufficiently removed from our current situation in order to be interesting, while nevertheless close enough to our core competence in order to be feasible. The SIT thinking process uses the FFF procedure and the five thinking tools in order to generate new forms (virtual products). These forms if unrestrained might lead to "far" ideas. The "closed world" condition for example acts as an inhibitor, making sure we don't wander too far out. This combination of opposite forces keeps us ideating in the sweet zone.

The Template Approach to New Product Development

SIT approaches New Product Development by identifying and applying certain well-defined schemes derived from an historical analysis of product-based trends, termed patterns or templates. These templates may contribute to the understanding and prediction of new product emergence. The invention of new products has traditionally involved methods that encourage the generation of large numbers of ideas. The notion that the rewards of generating a large number of ideas outweigh the costs can be traced back to early studies in the field. In view of the fact that this process tends to be highly complex and un-formalized, those involved in generating new ideas may seek ways to become more productive as they progress from one ideation task to another. Some may succeed in identifying patterns of invention that are common to different contexts and apply them within a certain product category, or even try to apply them to other product categories. Individuals who adopt such a cognitive strategy may expect to gain an advantage over others who treat every task as new and unrelated to past ideation. However, even if they prove productive, the patterns are likely to be idiosyncratic and, quite often, not even verbally definable. As such, they lack permanence and generalizability. SIT sets is built on the thesis that certain patterns are identifiable, objectively verifiable, widely applied, and learnable, and that these patterns, termed templates, can serve as a facilitative tool that channels the ideation process, enabling the individual to be more productive and focused.

The usual process of developing new products begins with a definition of market need. This is done based on intuition or on market analysis, focus groups etc. After defining the needs, a process is begun of developing products to address these needs. This process is called form follows function, as the form of the new product is derived from the function that it should fulfill. This process has a few disadvantages:

1. Most customers have difficulties in thinking about needs or products, which do not exist. This is particularly true for needs which are not vital. For example: how many customers thought of the need for a compact, transportable cassette player as addressed by Walkman? How many customers thought about the possibility to use the Internet communication as a means to conduct telephone calls?

2. To find those customers which do think of new needs\products, huge and very expensive surveys are needed. But even if you succeed in finding those people, chances are they will not be keen to share their good ideas for free.

3. If the need is clear or easy to define, it is most likely that, at least a few of your competitors have already defined it and are in the process of addressing the need.

In order to overcome these problems, the SIT method suggests starting the process of product development from the product itself. Applying systematic thinking tools in analyzing the product can lead to potential new products or to a definition of new needs. The advantages of this method are as follows:

1. The process requires only a limited amount of hours and is conducted in-house;

2. Applying the method yields many new ideas and a definition of many potential new needs;

3. As the new products are based on the old one, no major changes are usually required in production.

One of the important elements of SIT is to characterize the system and environmental variables. After having defined these variables, the participants are asked to examine the correlation between them, and to examine the impact of manipulating one or more of the product variables on the potential use of the "new" product; how such a change affects the correlation between the product and the environment and who may want to use such a product.

SIT in Universities and Business Schools

SIT is taught in numerous Universities and Business Schools worldwide. The methodology is most often taught through programs on Innovation; Business Administration; Marketing; Organizational Development; Leadership; Management Studies.

(ordered alphabetically)

- Bar Ilan University
- Ben-Gurion University of the Negev
- Brigham Young University
- Columbia University
- Duke University
- Hebrew University of Jerusalem
- INSEAD
- London Business School
- National University of Singapore
- Tel Aviv University
- Universidad de Los Andes Bogota
- University of Cincinnati
- University of North Carolina at Charlotte
- Wharton Business School

Brainstorming

Brainstorming is a group creativity technique by which efforts are made to find a conclusion for a specific problem by gathering a list of ideas spontaneously contributed by its members.

The term was popularized by Alex Faickney Osborn in the 1953 book *Applied Imagination.*

Origin

Advertising executive Alex F. Osborn began developing methods for creative problem-solving in 1939. He was frustrated by employees' inability to develop creative ideas individually for ad campaigns. In response, he began hosting group-thinking sessions and discovered a significant improvement in the quality and quantity of ideas produced by employees. Osborn outlined his method in the 1948 book *Your Creative Power* in chapter 33, "How to Organize a Squad to Create Ideas".

Osborn's Method

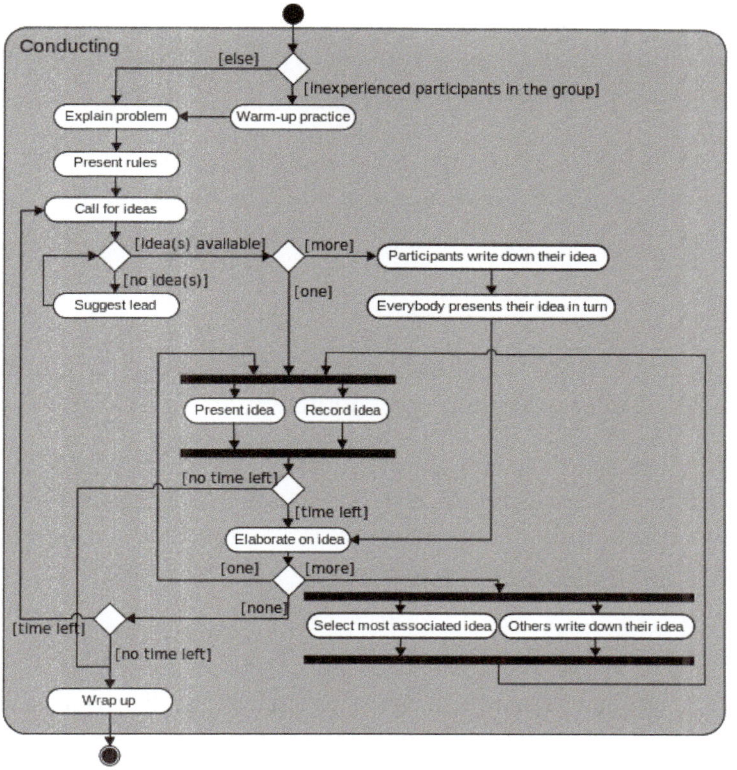

brainstorming activity conducting

Osborn claimed that two principles contribute to "ideative efficacy," these being :

1. Defer judgment,
2. Reach for quantity.

Following these two principles were his four general rules of brainstorming, established with intention to :

- reduce social inhibitions among group members.
- stimulate idea generation.
- increase overall creativity of the group.

1. go for quantity: This rule is a means of enhancing divergent production, aiming to facilitate problem solving through the maxim *quantity breeds quality*. The assumption is that the greater the number of ideas generated, the bigger the chance of producing a radical and effective solution.
2. Withhold criticism: In brainstorming, criticism of ideas generated should be put 'on hold'. Instead, participants should focus on extending or adding to

ideas, reserving criticism for a later 'critical stage' of the process. By suspending judgment, participants will feel free to generate unusual ideas.

3. Welcome wild ideas: To get a good long list of suggestions, wild ideas are encouraged. They can be generated by looking from new perspectives and suspending assumptions. These new ways of thinking might give you better solutions.
4. Combine and improve ideas:As suggested by the slogan "1+1=3". It is believed to stimulate the building of ideas by a process of association.

Applications

Osborn notes that brainstorming should address a specific question; he held that sessions addressing multiple questions were inefficient.

Further, the problem must require the generation of ideas rather than judgment; he uses examples such as generating possible names for a product as proper brainstorming material, whereas analytical judgments such as whether or not to marry do not have any need for brainstorming.

Groups

Osborn envisioned groups of around 12 participants, including both experts and novices. Participants are encouraged to provide wild and unexpected answers. Ideas receive no criticism or discussion. The group simply provides ideas that might lead to a solution and apply no analytical judgment as to the feasibility. The judgments are reserved for a later date.

Variations

Nominal Group Technique

Participants are asked to write their ideas anonymously. Then the facilitator collects the ideas and the group votes on each idea. The vote can be as simple as a show of hands in favor of a given idea. This process is called distillation.

After distillation, the top ranked ideas may be sent back to the group or to subgroups for further brainstorming. For example, one group may work on the color required in a product. Another group may work on the size, and so forth. Each group will come back to the whole group for ranking the listed ideas. Sometimes ideas that were previously dropped may be brought forward again once the group has re-evaluated the ideas.

It is important that the facilitator be trained in this process before attempting to facilitate this technique. The group should be primed and encouraged to embrace the process. Like all team efforts, it may take a few practice sessions to train the team in the method before tackling the important ideas.

Group Passing Technique

Each person in a circular group writes down one idea, and then passes the piece of paper to the next person, who adds some thoughts. This continues until everybody gets his or her original piece of paper back. By this time, it is likely that the group will have extensively elaborated on each idea.

The group may also create an "idea book" and post a distribution list or routing slip to the front of the book. On the first page is a description of the problem. The first person to receive the book lists his or her ideas and then routes the book to the next person on the distribution list. The second person can log new ideas or add to the ideas of the previous person. This continues until the distribution list is exhausted. A follow-up "read out" meeting is then held to discuss the ideas logged in the book. This technique takes longer, but it allows individuals time to think deeply about the problem.

Team Idea Mapping Method

This method of brainstorming works by the method of association. It may improve collaboration and increase the quantity of ideas, and is designed so that all attendees participate and no ideas are rejected.

The process begins with a well-defined topic. Each participant brainstorms individually, then all the ideas are merged onto one large idea map. During this consolidation phase, participants may discover a common understanding of the issues as they share the meanings behind their ideas. During this sharing, new ideas may arise by the association, and they are added to the map as well. Once all the ideas are captured, the group can prioritize and/or take action.

Directed Brainstorming

Directed brainstorming is a variation of electronic brainstorming (described below). It can be done manually or with computers. Directed brainstorming works when the solution space (that is, the set of criteria for evaluating a good idea) is known prior to the session. If known, those criteria can be used to constrain the Ideation process intentionally.

In directed brainstorming, each participant is given one sheet of paper (or electronic form) and told the brainstorming question. They are asked to produce one response and stop, then all of the papers (or forms) are randomly swapped among the participants. The participants are asked to look at the idea they received and to create a new idea that improves on that idea based on the initial criteria. The forms are then swapped again and respondents are asked to improve upon the ideas, and the process is repeated for three or more rounds.

In the laboratory, directed brainstorming has been found to almost triple the productivity of groups over electronic brainstorming.

Guided Brainstorming

A guided brainstorming session is time set aside to brainstorm either individually or as a collective group about a particular subject under the constraints of perspective and time. This type of brainstorming removes all cause for conflict and constrains conversations while stimulating critical and creative thinking in an engaging, balanced environment.

Participants are asked to adopt different mindsets for pre-defined period of time while contributing their ideas to a central mind map drawn by a pre-appointed scribe. Having examined a multi-perspective point of view, participants seemingly see the simple solutions that collectively create greater growth. Action is assigned individually.

Following a guided brainstorming session participants emerge with ideas ranked for further brainstorming, research and questions remaining unanswered and a prioritized, assigned, actionable list that leaves everyone with a clear understanding of what needs to happen next and the ability to visualize the combined future focus and greater goals of the group.

Individual Brainstorming

"Individual brainstorming" is the use of brainstorming in solitary situations. It typically includes such techniques as free writing, free speaking, word association, and drawing a mind map, which is a visual note taking technique in which people diagram their thoughts. Individual brainstorming is a useful method in creative writing and has been shown to be superior to traditional group brainstorming.

Question Brainstorming

This process involves brainstorming the *questions*, rather than trying to come up with immediate answers and short term solutions. Theoretically, this technique should not inhibit participation as there is no need to provide solutions. The answers to the questions form the framework for constructing future action plans. Once the list of questions is set, it may be necessary to prioritize them to reach to the best solution in an orderly way.

"Questorming" is another term for this mode of inquiry.

Electronic Brainstorming (EBS)

Although the brainstorming can take place online through commonly available technologies such as email or interactive web sites, there have also been many efforts to develop customized computer software that can either replace or enhance one or more manual elements of the brainstorming process.

Early efforts, such as GroupSystems at University of Arizona or Software Aided Meeting Management (SAMM) system at the University of Minnesota, took advantage of then-new computer networking technology, which was installed in rooms dedicated to computer supported meetings. When using these electronic meeting systems (EMS, as they came to be called), group members simultaneously and independently entered ideas into a computer terminal. The software collected (or "pools") the ideas into a list, which could be displayed on a central projection screen (anonymized if desired). Other elements of these EMSs could support additional activities such as categorization of ideas, elimination of duplicates, assessment and discussion of prioritized or controversial ideas. Later EMSs capitalized on advances in computer networking and internet protocols to support asynchronous brainstorming sessions over extended periods of time and in multiple locations

Introduced along with the EMS by Nunamaker and colleagues at University of Arizona was electronic brainstorming (EBS). By utilizing customized computer software for groups (*group decision support systems* or *groupware*), EBS can replace face-to-face brainstorming. An example of groupware is the *GroupSystems,* a software developed by University of Arizona. After an idea discussion has been posted on *GroupSystems,* it is displayed on each group member's computer. As group members simultaneously type their comments on separate computers, those comments are anonymously pooled and made available to all group members for evaluation and further elaboration.

Compared to face-to-face brainstorming, not only does EBS enhanced efficiency by eliminating travelling and turn-taking during group discussions, it also excluded several psychological constraints associated with face-to-face meetings. Identified by Gallupe and colleagues, both production blocking (reduced idea generation due to turn-taking and forgetting ideas in face-to-face brainstorming) and evaluation apprehension (a general concern experienced by individuals for how others in the presence are evaluating them) are reduced in EBS. These positive psychological effects increase with group size. A perceived advantage of EBS is that all ideas can be archived electronically in their original form, and then retrieved later for further thought and discussion. EBS also enables much larger groups to brainstorm on a topic than would normally be productive in a traditional brainstorming session.

Computer supported brainstorming may overcome some of the challenges faced by traditional brainstorming methods. For example, ideas might be "pooled" automatically, so that individuals do not need to wait to take a turn, as in verbal brainstorming. Some software programs show all ideas as they are generated (via chat room or e-mail). The display of ideas may cognitively stimulate brainstormers, as their attention is kept on the flow of ideas being generated without the potential distraction of social cues such as facial expressions and verbal language. EBS techniques have been shown to produce more ideas and help individuals focus their attention on the ideas of others better than a brainwriting technique (participants write individual written notes in silence and then subsequently communicate them with the group). The production of more ideas has been linked to the fact that paying attention to others' ideas leads to non-redundancy, as brainstormers try to avoid to replicate or repeat another participant's comment or idea. Conversely, the production gain associated with EBS was less found in situations where EBS group members focused too much on generating ideas that they ignored ideas expressed by others. The production gain associated with *GroupSystem* users' attentiveness to ideas expressed by others has been documented by Dugosh and colleagues. EBS group members who were instructed to attend to ideas generated by others outperformed those who were not in terms of creativity.

According to a meta-analysis comparing EBS to face-to-face brainstorming conducted by DeRosa and colleagues, EBS has been found to enhance both the production of non-redundant ideas and the quality of ideas produced. Despite the advantages demonstrated by EBS groups, EBS group members reported less satisfaction with the brainstorming process compared to face-to-face brainstorming group members.

Some web-based brainstorming techniques allow contributors to post their comments anonymously through the use of avatars. This technique also allows users to log on over an extended time period, typically one or two weeks, to allow participants some "soak time" before posting their ideas and feedback. This technique has been used particularly in the field of new product development, but can be applied in any number of areas requiring collection and evaluation of ideas.

Some limitations of EBS include the fact that it can flood people with too many ideas at one time that they have to attend to, and people may also compare their performance to others by analyzing how many ideas each individual produces (social matching).

Incentives

Some research indicates that incentives can augment creative processes. Participants were divided into three conditions. In Condition I, a flat fee was paid to all participants. In the Condition II, participants were awarded points for every unique idea of their own, and subjects were paid for the points that they earned. In Condition III, subjects were paid based on the impact that their idea had on the group; this was measured by counting the number of group ideas derived from the specific subject's ideas. Condition III outperformed Condition II, and Condition II outperformed Condition I at a statistically significant level for most measures. The results demonstrated that participants were willing to work far longer to achieve unique results in the expectation of compensation.

Challenges to Effective Group Brainstorming

A good deal of research refutes Osborn's claim that group brainstorming could generate more ideas than individuals working alone. For example, in a review of 22 studies of group brainstorming, Michael Diehl and Wolfgang Stroebe found that, overwhelmingly, groups brainstorming together produce fewer ideas than individuals working separately. However, this conclusion is brought into question by a subsequent review of 50 studies by Scott G. Isaksen showed that a misunderstanding of the tool, and weak application of the methods (including lack of facilitation), and the artificiality of the problems and groups undermined most such studies, and the validity of their conclusions.

Several factors can contribute to a loss of effectiveness in group brainstorming.

Blocking:

Because only one participant may give an idea at any one time, other participants might forget the idea they were going to contribute or not share it because they see it as no longer important or relevant. Further, if we view brainstorming as a cognitive process in which "a participant generates ideas (generation process) and stores them in short-term memory (memorization process) and then eventually extracts some of them from its short-term memory to express them (output process)", then blocking is an even more critical challenge because it may also inhibit a person's train of thought in generating their own ideas and remembering them.

Collaborative fixation: Exchanging ideas in a group may reduce the number of domains that a group explores for additional ideas. Members may also conform their ideas to those of other members, decreasing the novelty or variety of ideas, even though the overall number of ideas might not decrease.

Evaluation apprehension: Evaluation apprehension was determined to occur only in instances of personal evaluation. If the assumption of collective assessment were in place, real-time judgment of ideas, ostensibly an induction of evaluation apprehension, failed to induce significant variance.

Free-riding: Individuals may feel that their ideas are less valuable when combined with the ideas of the group at large. Indeed, Diehl and Stroebe demonstrated that even when individuals worked alone, they produced fewer ideas if told that their output would be judged in a group with others than if told that their output would be judged individually. However, experimentation revealed free riding as only a marginal contributor to productivity loss, and type of session (i.e., real vs. nominal group) contributed much more.

Personality characteristics: Extroverts have been shown to outperform introverts in computer mediated groups. Extroverts also generated more unique and diverse ideas than introverts when additional methods were used to stimulate idea generation, such as completing a small related task before brainstorming, or being given a list of the classic rules of brainstorming.

Social matching: One phenomenon of group brainstorming is that participants will tend to alter their rate of productivity to match others in the group. This can lead to participants generating fewer ideas in a group setting than they would individually because they will decrease their own contributions if they perceive themselves to be more productive than the group average. On the other hand, the same phenomenon can also increase an individual's rate of production to meet the group average.

Nominal Group Technique

The nominal group technique (NGT) is a group process involving problem identification, solution generation, and decision making. It can be used in groups of many sizes, who want to make their decision quickly, as by a vote, but want everyone's opinions taken into account (as opposed to traditional voting, where only the largest group is considered). The method of tallying is the difference. First, every member of the group gives their view of the solution, with a short explanation. Then, duplicate solutions are eliminated from the list of all solutions, and the members proceed to rank the solutions, 1st, 2nd, 3rd, 4th, and so on.

Some facilitators will encourage the sharing and discussion of reasons for the choices made by each group member, thereby identifying common ground, and a plurality of ideas and approaches. This diversity often allows the creation of a hybrid idea (combining parts of two or more ideas), often found to be even better than those ideas being initially considered.

In the basic method, the numbers each solution receives are totaled, and the solution with the highest (i.e. most favored) total ranking is selected as the final decision. There are variations on how this technique is used. For example, it can identify strengths versus areas in need of development, rather than be used as a decision-making voting alternative. Also, options do not always have to be ranked, but may be evaluated more subjectively.

This technique was originally developed by Andre Delbecq and Andrew H. Van de Ven, and has been applied to adult education program planning by Vedros, and has also been employed as a useful technique in curriculum design and evaluation in educational institutions

Effects

NGT have been shown to enhance one or more dimensions of effectiveness of decision-making groups. Requiring individuals to write down their ideas silently and independently prior to a group discussion increased the number of solutions generated by groups. Round-robin polling also resulted in a larger number of inputs and fostered more equal participation. The increased number of heterogeneous inputs led to high quality decisions.

As compared to interacting groups the NGT groups provide more unique ideas, more balanced participation between group members, increased feelings of accomplishment, and greater satisfaction with idea quality and group efficiency.

These findings are consistent with a 1958 study which found that, in response to three different problems requiring creative thinking, the number of ideas produced by "nominal groups" (whose members were actually working alone) was greater than the number of ideas produced by real, face-to-face groups. The ideas generated by the nominal and real groups were rated qualitatively and for originality, and the nominal groups scored better on both of those measures.

Usage

The nominal group technique is particularly useful:

- When some group members are much more vocal than others.
- When some group members think better in silence.
- When there is concern about some members not participating.
- When the group does not easily generate quantities of ideas.
- When all or some group members are new to the team.
- When the issue is controversial or there is heated conflict.
- When there is a power-imbalance between facilitator and participants or participants: the structure of the NGT session can balance these out.
- When stakeholders like a(/some) quantitative output of the process.

Standard Procedure

Routinely, the NGT involves five stages:

1. Introduction and explanation: The facilitator welcomes the participants and explains to them the purpose and procedure of the meeting.
2. Silent generation of ideas: The Facilitator provides each participant with a sheet of paper with the question to be addressed and ask them to write down all ideas that come to mind when considering the question. During this period, the facilitator asks participants not to consult or discuss their ideas with others. This stage lasts approximately 10 minutes.
3. Sharing ideas: The Facilitator invites participants to share the ideas they have generated. He records each idea on a flip chart using the words spoken by the participant. The round robin process continues until all ideas have been presented. There is no debate about items at this stage and participants are encouraged to write down any new ideas that may arise from what others share. This process ensures all participants get an opportunity to make an equal contribu-

tion and provides a written record of all ideas generated by the group. This stage may take 15–30 minutes.

4. Group discussion: Participants are invited to seek verbal explanation or further details about any of the ideas that colleagues have produced that may not be clear to them. The facilitator's task is to ensure that each person is allowed to contribute and that discussion of all ideas is thorough without spending too long on a single idea. It is important to ensure that the process is as neutral as possible, avoiding judgment and criticism. The group may suggest new items for discussion and combine items into categories, but no ideas should be eliminated. This stage lasts 30–45 minutes.

5. Voting and ranking: This involves prioritizing the recorded ideas in relation to the original question. Following the voting and ranking process, immediate results in response to the question is available to participants so the meeting concludes having reached a specific outcome.

The number of nominal group meetings to be held will depend on the nature of the question and accessibility to the key stakeholders best suited to help address the problem.

Advantages and Disadvantages

One major advantage of NGT is that it avoids two problems caused by group interaction. First, some members are reluctant to suggest ideas because they are concerned about being criticized, or are reticent and shy. Second, some members are reluctant to create conflict in groups. (Many people want to maintain a pleasant climate.) NGT overcomes these problems (e.g.). NGT has the clear advantage in ensuring relatively equal participation. It may also, in many cases be a time-saving technique. Other advantages include producing a large number of ideas and providing a sense of closure that is often not found in less-structured group methods.

A major disadvantage of NGT is that the method lacks flexibility by being able to deal with only one problem at a time. Also, there must be a certain amount of conformity on the part of the members involved in NGT. Everyone must feel comfortable with the amount of structure involved. Another disadvantage is the amount of time needed to prepare for the activity. There is no spontaneity involved with this method. Facilities must be arranged and carefully planned. Opinions may not converge in the voting process, cross-fertilization of ideas may be constrained, and the process may appear to be too mechanical.

One of the key issues about 'nominal' group technique is that it does not depend on normal group processes. It is a method to work with a collection of people and involve them in decision making but does not depend on existing group processes. This is according to the originators an advantage in decision making using this tool.

Adaptation for Ill-structured Problems

Modification of NGT, undertaken by Bartunek and Murnighan , helps to deal with ill-structured problems. Normal ideas are generated and listed, followed by the facilitator questioning if the ideas are relevant to the same problem. If not, the problem is said to be ill-structured, and the ideas generated are clustered into coherent groups. These clusters of ill-structured ideas are then treated as problems in their own right, and the NGT procedure is applied to them. Regular breaks are taken by the participants to ensure that the group feels they are still working on the original problem.

Eight Disciplines Problem Solving

Eight Disciplines (8Ds) Problem Solving is a method developed at Ford Motor Company used to approach and to resolve problems, typically employed by engineers or other professionals. Focused on product and process improvement, its purpose is to identify, correct, and eliminate recurring problems. It establishes a permanent corrective action based on statistical analysis of the problem and on the origin of the problem by determining the root causes. Although it originally comprised eight stages, or 'disciplines', it was later augmented by an initial planning stage. 8D follows the logic of the PDCA cycle. The disciplines are:

D0: Plan: Plan for solving the problem and determine the prerequisites.

D1: Use a Team: Establish a team of people with product/process knowledge.

D2: Describe the Problem: Specify the problem by identifying in quantifiable terms the who, what, where, when, why, how, and how many (5W2H) for the problem.

D3: Develop Interim Containment Plan: Define and implement containment actions to isolate the problem from any customer.

D4: Determine, and Verify Root Causes and Escape Points: Identify all applicable causes that could explain why the problem has occurred. Also identify why the problem was not noticed at the time it occurred. All causes shall be verified or proved. One can use five whys or Ishikawa diagrams to map causes against the effect or problem identified.

D5: Verify Permanent Corrections (PCs) for Problem will resolve problem for the customer: Using pre-production programs, quantitatively confirm that the selected correction will resolve the problem. (Verify that the correction will actually solve the problem.)

D6: Define and Implement Corrective Actions: Define and Implement the best corrective actions.

D7: Prevent System Problems: Modify the management systems, operation systems, practices, and procedures to prevent recurrence of this and all similar problems.

D8: Congratulate Your Team: Recognize the collective efforts of the team. The team needs to be formally thanked by the organization.

8Ds has become a standard in the automotive, assembly, and other industries that require a thorough structured problem solving process using a team approach.

Ford Motor Company's Team Oriented Problem Solving TOPS

The executives of the Powertrain Organization (transmissions, chassis, engines) wanted a methodology where teams (design engineering, manufacturing engineering, and production) could work on recurring chronic problems. In 1986, the assignment was given to develop a manual and a subsequent course that would achieve a new approach to solving identified engineering design and manufacturing problems. The manual for this methodology was documented and defined in *Team Oriented Problem Solving* (TOPS), first published in 1987. The manual and subsequent course material were piloted at Ford World Headquarters in Dearborn, Michigan. This has been Ford's established approach to problem solving. This methodology included good practices associated with eliminating engineering product problems. Ford refers to their current variant as G8D (Global 8D). The Ford 8Ds manual is extensive and covers chapter by chapter how to go about addressing, quantifying, and resolving engineering issues. It begins with a cross functional team and concludes with a successful demonstrated resolution of the problem. Containment actions may or may not be needed based on where the problem occurred in the life cycle of the product.

Usage

Many disciplines are typically involved in the "8Ds" methodology. The tools used can be found in textbooks and reference materials used by quality assurance professionals. For example, an "Is/Is Not" worksheet is a common tool employed at D2, and Ishikawa, or "fishbone," diagrams and "5-why analysis" are common tools employed at step D4. In the late 1990s, Ford developed a revised version of the 8D process that they call "Global 8D" (G8D), which is the current global standard for Ford and many other companies in the automotive supply chain. The major revisions to the process are as follows:

- Addition of a D0 (D-Zero) step as a gateway to the process. At D0, the team documents the symptoms that initiated the effort along with any emergency response actions (ERAs) that were taken before formal initiation of the G8D. D0 also incorporates standard assessing questions meant to determine whether a full G8D is required. The assessing questions are meant to ensure that in a world of limited problem-solving resources, the efforts required for a full team-based problem-solving effort are limited to those problems that warrant these resources.

- Addition of the notion of escape points to D4 through D6. An 'escape point' is the earliest control point in the control system following the root cause of a problem that should have detected that problem but failed to do so. The idea here is to consider not only the root cause, but also what went wrong with the control system in allowing this problem to escape. Global 8D requires the team to identify and verify an escape point at D4. Then, through D5 and D6, the process requires the team to choose, verify, implement, and validate permanent corrective actions to address the escape point.

Recently, the 8D process has been employed significantly outside the auto industry. As part of lean initiatives and continuous-improvement processes it is employed extensively in the food manufacturing, health care, and high-tech manufacturing industries.

Benefits

The benefits of the 8D methodology include effective approaches to finding a root cause, developing proper actions to eliminate root causes, and implementing the permanent corrective action. The system also helps to explore the control systems that allowed the problem to escape. The Escape Point is studied for the purpose of improving the ability of the Control System to detect the failure or cause when and if it should occur again. Finally the Prevention Loop explores the systems that permitted the condition that allowed the Failure and Cause Mechanism to exist in the first place.

Disadvantages

Requires training in the 8D problem-solving process as well as appropriate data collection and analysis tools such as Pareto charts, Fishbone Diagrams, and process maps.

Problem Solving Tools

The following tools can be used within 8D:

- Ishikawa diagrams also known as Cause and Effect or Fishbone diagrams
- Pareto charts or Pareto diagrams
- 5 Whys root cause analysis
- 5W and 2H (Who, What, Where, When, Why, How, How often)
- Statistical Process Control
- Scatter plots

- Design of Experiments
- Check sheet
- Histograms
- FMEA
- Flowcharts or Process Maps

Background of Common Corrective Actions to Dispose of Nonconforming Items

The 8D methodology was first described in a Ford manual in 1987. The manual describes the eight-step methodology to address chronic product and process problems. The 8Ds included several concepts of effective problem solving, including taking corrective actions and containing nonconforming items. These two steps have been very common in most manufacturing facilities, including government and military installations. In 1974, the U.S. Department of Defense (DOD) released "MIL-STD 1520 Corrective Action and Disposition System for Nonconforming Material". This 13 page standard defines establishing some corrective actions and then taking containment actions on nonconforming material or items. It is focused on inspection for defects and disposing of them. The basic idea of corrective actions and containment of defectives was officially abolished in 1995, but these concepts were also common to Ford Motor Company, a major supplier to the government in World War II. Corrective actions and containment of poor quality parts were part of the manual and course for the automotive industry and are well known to many companies. Ford's 60 page manual covers details associated with each step in their 8D problem solving manual and the actions to take to deal with identified problems.

Military usage

The US Government standardized a procedure during the Second World War to address nonconforming material in Military Standard 1520, *Corrective Action and Disposition System for Nonconforming Material*. This military standard focused on nonconforming material and the disposition of the material.

Relationship Between 8D and FMEA

FMEA (Failure Mode and Effect Analysis) is a tool generally used in the planning of product or process design. The relationships between 8D and FMEA are outlined below:

1. The problem statements and descriptions are sometimes linked between both documents. An 8D can utilize pre-brainstormed information from a FMEA to assist in looking for potential problems.

2. Possible causes in a FMEA can immediately be used to jump start 8D Fishbone or Ishikawa diagrams. Brainstorming information that is already known is not a good use of time or resources.

3. Data and brainstorming collected during an 8D can be placed into a FMEA for future planning of new product or process quality. This allows a FMEA to consider actual failures, occurring as failure modes and causes, becoming more effective and complete.

4. The design or process controls in a FMEA can be used in verifying the root cause and Permanent Corrective Action in an 8D.

The FMEA and 8D should reconcile each failure and cause by cross documenting failure modes, problem statements and possible causes. Each FMEA can be used as a database of possible causes of failure as an 8D is developed.

GROW Model

The GROW model (or process) is a simple method for goal setting and problem solving. It was developed in the United Kingdom and was used extensively in corporate coaching in the late 1980s and 1990s.

There have been many claims to authorship of GROW as a way of achieving goals and solving problems. While no one person can be clearly identified as the originator, Graham Alexander, Alan Fine, and Sir John Whitmore all made significant contributions. Max Landsberg also describes GROW in his book *The Tao of Coaching*. Other similar models include *collaborative helping maps* in family therapy and Gabriele Oettingen's WOOP model.

Jonathan Passmore and Stefan Cantore have suggested that one 'argument against behavioural-based approaches such as GROW is that their goal nature excludes the potential to explore philosophical aspects of life. Thus GROW may be suited to working in goal-directed areas of sports or business, but may be less well suited to careers conversations, person–role fit or life-coaching conversations where other approaches such as the transpersonal or existential approaches may be more helpful.'

Stages of GROW

There are a number of different versions of the GROW model. The following table presents one view of the stages but there are others. The 'O' in this version has two meanings.

G	Goal	The Goal is the end point, where the client wants to be. The goal has to be defined in such a way that it is very clear to the client when they have achieved it.

R	Reality	The Current Reality is where the client is now. What are the issues, the challenges, how far are they away from their goal?
O	Obstacles	There will be Obstacles stopping the client getting from where they are now to where they want to go. If there were no Obstacles the client would already have reached their goal.
	Options	Once Obstacles have been identified, the client needs to find ways of dealing with them if they are to make progress. These are the Options.
W	Way Forward	The Options then need to be converted into action steps which will take the client to their goal. These are the Way Forward.

As with many simple principles, any user of GROW can apply a great deal of skill and knowledge at each stage but the basic process remains as written above. There are numerous questions which the coach could use at any point and part of the skill of the coach is to know which questions to use and how much detail to uncover.

Example

The following is a very simple example of using the GROW model to achieve a goal. This example deals with weight loss. If the client wants: 'To bring my weight down to 120 pounds in three months and keep it down', that is their *Goal*. The more heartfelt and personal, the more meaningful the goal is to the person and the more likely they will be to commit to and achieve the goal.

The GROW approach would then be to establish the *Reality* by stating what their weight is now. The coach would then ask awareness questions to deepen understanding of what is happening when the client tries to lose weight, thus identifying the *Obstacles*. These questions could include:

- When you have been able to lose weight — what made the difference?
- What is the difference between the times you are able to keep weight off and the times when you put it on again?
- What would have to change for you to be sure you could lose the weight and keep it off?

If the client genuinely answers these questions they will discover new information about what works and does not work for them in terms of weight loss, and create some potential for change. It then becomes possible to create some strategies or *Options* which get around the *Obstacles*. These could include looking at which diets or exercise regimes work best, or finding a specific type of support. Once the client knows the strategies that are likely to work they can establish a *Way Forward* which involves taking action steps. This is where they commit to what they will do in the short term to put the strategies into effect. For instance, one action might be asking a particular person for support, and another might be to buy a different selection of foods.

GROW neatly highlights the nature of a problem for coaching purposes. In order for a problem to exist in coaching terms there has to be two elements present. Firstly there has to be something that the client is trying to achieve — the *Goal*. Then there has to be something stopping them achieve that goal — the *Obstacle(s)*. Using GROW automatically breaks a problem down into these component parts.

The same principles can be applied whatever goal or problem the client has. GROW can be used on technical problems, issues regarding processes, strategy questions, interpersonal issues and many more. The model can also be used by a group who are all working on the same problem or goal.

The GROW principle and the Inner Game

GROW was influenced by the Inner Game method developed by Timothy Gallwey. Gallwey was a tennis coach who noticed that he could often see what players were doing incorrectly but that simply telling them what they should be doing did not bring about lasting change.

The parallel between Gallwey's Inner Game method and the GROW method can be illustrated by the example of players who do not keep their eyes on the ball. Some coaches might give instructions such as: 'Keep your eye on the ball' to try to correct this. The problem with this sort of instruction is that a player will be able to follow it for a short while but may be unable to keep it in mind in the long term. So one day, instead of giving an instruction, Gallwey asked players to say 'bounce' out loud when the ball bounced and 'hit' out loud when they hit the ball.

The result was that the players started to improve without a lot of effort because they were keeping their eyes on the ball. But because of the way the instruction was given they did not have a voice in their heads saying 'I must keep my eye on the ball.' Instead they were playing a simple game while they were playing tennis. Once Gallwey saw how play could be improved in this way, he stopped giving instructions and started asking questions that would help players discover for themselves what worked and what needed to change.

The GROW method is similar. For example, the first stage in the learning process would be to set a target which a player wants to achieve. If a player wanted to improve their first serve Gallwey would ask how many first serves out of ten they would like to get in. This is the *Goal*. The *Reality* would be defined by asking the player to serve 10 balls and seeing how many first serves went in.

Gallwey would then ask awareness-raising questions such as 'What do you notice you are doing differently when the ball goes in or out?' This question would enable players to discover for themselves what was changing about their mind and body when the serve went in or out. They had then defined their *Obstacles* and *Options*. They therefore learned for themselves what had to change in order to meet their serving targets and they had a clear *Way Forward*.

The originators of both the Inner Game method and the GROW method suggested that many individuals were struggling to achieve goals because they were not learning from experience and were not aware of the available knowledge that would help them.

TRIZ

TRIZ (/ˈtriːz/; Russian: теория решения изобретательских задач, *teoriya resheniya izobretatelskikh zadach*, literally: "theory of the resolution of invention-related tasks") is "a problem-solving, analysis and forecasting tool derived from the study of patterns of invention in the global patent literature". It was developed by the Soviet inventor and science-fiction author Genrich Altshuller (1926-1998) and his colleagues, beginning in 1946. In English the name is typically rendered as "the theory of inventive problem solving", and occasionally goes by the English acronym TIPS.

Following Altshuller's insight, the theory developed on a foundation of extensive research covering hundreds of thousands of inventions across many different fields to produce a theory which defines generalisable patterns in the nature of inventive solutions and the distinguishing characteristics of the problems that these inventions have overcome.

An important part of the theory has been devoted to revealing patterns of evolution and one of the objectives which has been pursued by leading practitioners of TRIZ has been the development of an algorithmic approach to the invention of new systems, and to the refinement of existing ones.

TRIZ includes a practical methodology, tool sets, a knowledge base, and model-based technology for generating innovative solutions for problem solving. It is intended for application in problem formulation, system analysis, failure analysis, and patterns of system evolution. There is a general similarity of purposes and methods with the field of pattern language, a cross discipline practice for explicitly describing and sharing holistic patterns of design.

The research has produced three primary findings:

1. problems and solutions are repeated across industries and sciences
2. patterns of technical evolution are also repeated across industries and sciences
3. the innovations used scientific effects outside the field in which they were developed

TRIZ practitioners apply all these findings in order to create and to improve products, services, and systems.

History

TRIZ in its classical form was developed by the Soviet inventor and science fiction writer Genrich Altshuller and his associates. He started developing TRIZ in 1946 while working in the "Inventions Inspection" department of the Caspian Sea flotilla of the Soviet Navy. His job was to help with the initiation of invention proposals, to rectify and document them, and to prepare applications to the patent office. During this time he realised that a problem requires an inventive solution if there is an unresolved contradiction in the sense that improving one parameter impacts negatively on another. He later called these "technical contradictions".

His work on what later resulted in TRIZ was interrupted in 1950 by his arrest and sentencing to 25 years in the Vorkuta Gulag labor camps. The arrest was partially triggered by letters which he and Raphael Shapiro sent to Stalin, ministers and newspapers about certain decisions made by the Soviet Government, which they believed were erroneous. Altshuller and Shapiro were freed during the Khrushchev Thaw following Stalin's death in 1953 and returned to Baku.

The first paper on TRIZ titled "On the psychology of inventive creation" was published in 1956 in "Issues in Psychology" (Voprosi Psichologii) journal.

By 1969, Altshuller had reviewed about 40,000 patent abstracts in order to find out in what way the innovation had taken place and developed the concept of technical contradictions, the concept of ideality of a system, contradiction matrix, and 40 principles of invention. In the years that followed he developed the concepts of physical contradictions, SuField analysis (structural substance-field analysis), standard solutions, several laws of technical systems evolution, and numerous other theoretical and practical approaches.

Altshuller also observed clever and creative people at work: he uncovered patterns in their thinking, and developed thinking tools and techniques to model this "talented thinking". These tools include Smart Little People and Thinking in Time and Scale (or the Screens of Talented Thought).

In 1971 Altshuller convinced The Inventors Society to establish in Baku the first TRIZ teaching facility, called the Azerbaijan Public Institute for Inventive Creation and the first TRIZ research lab called The Public Lab for Inventive Creation. Altshuller was appointed the head of the lab by the society. The lab incubated the TRIZ movement and in the years that followed other TRIZ teaching institutes were established in all major cities of the USSR.

From 1986 Altshuller switched his attention away from technical TRIZ, and started investigating the development of individual creativity. He also developed a version of TRIZ for children, which was trialled in various schools. In 1989 the TRIZ Association was formed, with Altshuller chosen as President.

Following the end of the cold war, the waves of emigrants from the former Soviet Union brought TRIZ to other countries and drew attention to it overseas. In 1995 the Altshuller Institute for TRIZ Studies was established in Boston, USA.

Basic Principles

TRIZ presents a systematic approach for understanding and defining challenging problems: difficult problems require an inventive solution, and TRIZ provides a range of strategies and tools for finding these inventive solutions. One of the earliest findings of the massive research on which the theory is based is that the vast majority of problems that require inventive solutions typically reflect a need to overcome a dilemma or a trade-off between two contradictory elements. The central purpose of TRIZ-based analysis is to systematically apply the strategies and tools to find superior solutions that overcome the need for a compromise or trade-off between the two elements.

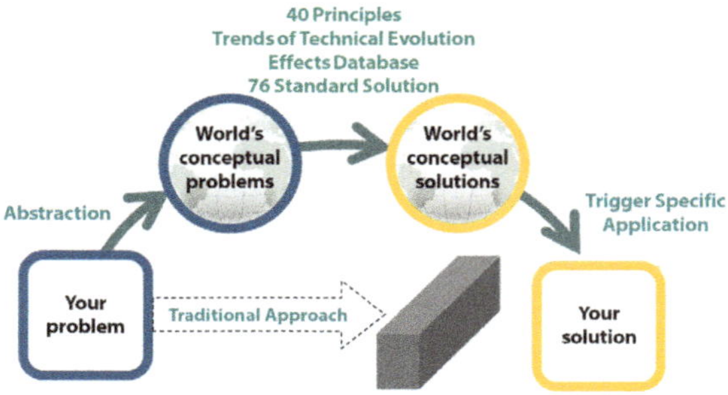

Prism of TRIZ

By the early 1970s two decades of research covering hundreds of thousands of patents had confirmed Altshuller's initial insight about the patterns of inventive solutions and one of the first analytical tools was published in the form of 40 inventive principles, which could account for virtually all of those patents that presented truly inventive solutions. Following this approach the "Conceptual solution" shown in the diagram can be found by defining the contradiction which needs to be resolved and systematically considering which of the 40 principles may be applied to provide a specific solution which will overcome the "contradiction" in the problem at hand, enabling a solution that is closer to the "ultimate ideal result".

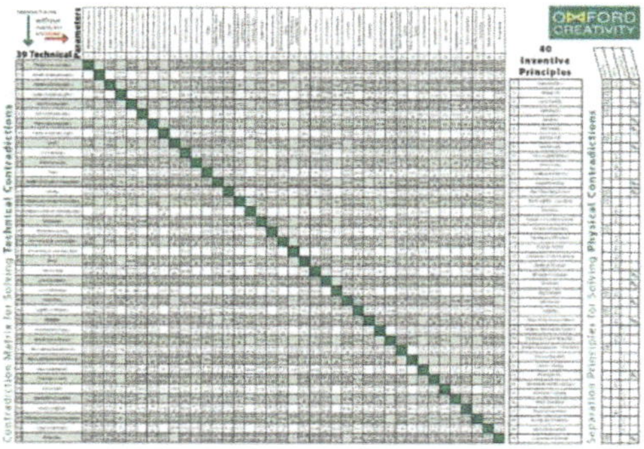

Contradictions Matrix

The combination of all of these concepts together – the analysis of the contradiction, the pursuit of an ideal solution and the search for one or more of the principles which will overcome the contradiction, are the key elements in a process which is designed to help the inventor to engage in the process with purposefulness and focus.

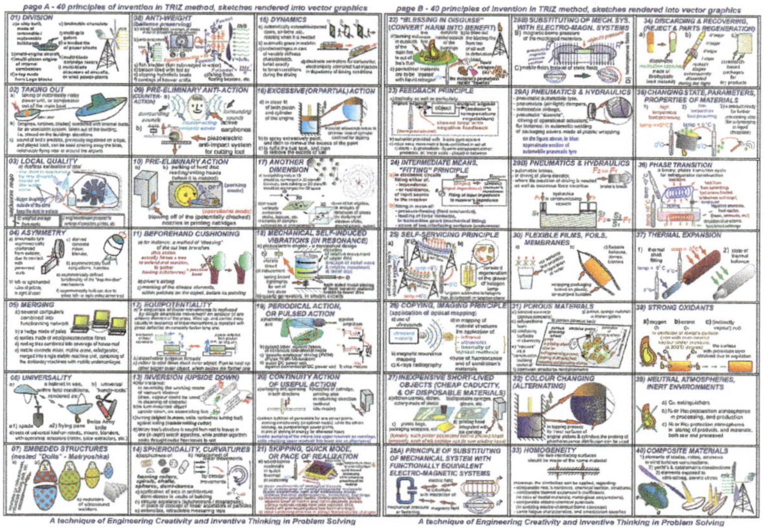

40 principles of TRIZ method rendered schematically

One of the tools which evolved as an extension of the 40 principles was a contradiction matrix in which the contradictory elements of a problem were categorized according to a list of 39 factors which could impact on each other. The combination of each pairing of these 39 elements is set out in a matrix (for example, the weight of a stationary object, the use of energy by a moving object, the ease of repair etc.) Each of the 39 elements is represented down the rows and across the columns (as the negatively affected element) and based upon the research and analysis of patents: wherever precedent solutions have been found that resolve a conflict between two of the elements, the relevant cells in the matrix typically contain a sub-set of three or four principles that have been applied most frequently in inventive solutions which resolve contradictions between those two elements.

The main objective of the contradiction matrix was to simplify the process of selecting the most appropriate Principle to resolve a specific contradiction. It was the core of all modifications of ARIZ till 1973. But in 1973, after introducing the concept of physical contradictions and creating SuField analysis, Altshuller realized that the contradiction matrix was comparatively an inefficient tool and stopped working on it. Beginning ARIZ-71c contradiction matrix ceased to be the core of ARIZ and therefore was not a tool for solving inventive problems that Altshuller believed should be pursued.

Physical contradictions and separation principles as well as SuField analysis, etc. became the core. Despite this, the 40 principles of invention has remained the most popular tool taught in introductory seminars and has consistently attracted the most attention amongst the tens of thousands of individuals who visit TRIZ-focused web sites in a typical month. Therefore, many of those who learn TRIZ or have attended seminars are taught quite wrongly that TRIZ is primarily composed of the 40 principles and contradiction matrix, the truth is ARIZ is the core methodology of TRIZ.

ARIZ is an algorithmic approach to finding inventive solutions by identifying and resolving contradictions. This includes the "system of inventive standards solutions" which Altshuller used to replace the 40 principles and contradiction matrix, it consists of SuField modeling and the 76 inventive standards. A number of TRIZ-based computer programs have been developed whose purpose is to provide assistance to engineers and inventors in finding inventive solutions for technological problems. Some of these programs are also designed to apply another TRIZ methodology whose purpose is to reveal and forecast emergency situations and to anticipate circumstances which could result in undesirable outcomes.

One of the important branches of TRIZ is focused on analysing and predicting trends of evolution in the characteristics that existing solutions are likely to develop in successive generations of a system.

Essentials

Basic Terms

- *Ideal final result* (IFR) - the ultimate idealistic solution of a problem when the desired result is achieved by itself. Note that the Ideal Final Result is also an ARIZ term for the formulation of the inventive problem in the form of a Technical Contradiction (IFR-1) and a Physical Contradiction (IFR-2);
- *Administrative contradiction* - contradiction between the needs and abilities;
- *Technical contradiction* - an inverse dependence between parameters/characteristics of a machine or technology;
- *Physical contradiction* - opposite/contradictory physical requirements to an object;
- *Separation principle* - a method of resolving physical contradictions by separating contradictory requirements;
- *Vepol* or *Su-field* - a minimal technical system consisting of two material objects (substances) and a "field". "Field" is the source of energy whereas one of the substances is "transmission" and the other one is the "tool";
- *Fepol* or *Ferfiel* - a sort of Vepol (Su-field) where "substances" are ferromagnetic objects;
- *Level of invention*;
- *Standard solution* - a standard inventive solution of a higher level;

- *Laws of technical systems evolution;*
- *Algorithm of inventive problems solving (ARIZ), which combines various specialized methods of TRIZ into one universal tool;*
- *Talented Thinking or Thinking in Time and Scale;*

Identifying a Problem: Contradictions

Altshuller has shown that at the heart of some inventive problems lie contradictions (one of the basic TRIZ concepts) between two or more elements, such as, "If we want more acceleration, we need a larger engine; but that will increase the cost of the car," that is, more of something desirable also brings more of something less desirable, or less of something else also desirable.

These are called *technical contradictions* by Altshuller. He also defined so-called physical or inherent contradictions: More of one thing and less of the same thing may both be desired in the same system. For instance, a higher temperature may be needed to melt a compound more rapidly, but a lower temperature may be needed to achieve a homogeneous mixture.

An *inventive situation* which challenges us to be inventive, might involve several such contradictions. Conventional solutions typically "trade" one contradictory parameter for another; no special inventiveness is needed for that. Rather, the inventor would develop a creative approach for resolving the contradiction, such as inventing an engine that produces more acceleration without increasing the cost of the engine.

Inventive Principles and the Matrix of Contradictions

Altshuller screened patents in order to find out what kind of contradictions were resolved or dissolved by the invention and the way this had been achieved. From this he developed a set of 40 inventive principles and later a matrix of contradictions. Rows of the matrix indicate the 39 system features that one typically wants to improve, such as speed, weight, accuracy of measurement and so on. Columns refer to typical undesired results. Each matrix cell points to principles that have been most frequently used in patents in order to resolve the contradiction.

For instance, Dolgashev mentions the following contradiction: increasing accuracy of measurement of machined balls while avoiding the use of expensive microscopes and elaborate control equipment. The matrix cell in row "accuracy of measurement" and column "complexity of control" points to several principles, among them the Copying Principle, which states, "Use a simple and inexpensive optical copy with a suitable scale instead of an object that is complex, expensive, fragile or inconvenient to operate." From this general invention principle, the following idea might solve the problem: Taking a high-resolution image of the machined ball. A screen with a grid might provide the required measurement. As mentioned above, Altshuller abandoned this method of defining and solving "technical" contradictions in the mid 1980s and instead used SuField modeling and the 76 inventive standards and a number of other tools included in the algorithm for solving inventive problems, ARIZ.

Laws of Technical System Evolution

Altshuller also studied the way technical systems have been developed and improved over time. From this, he discovered several trends (so called Laws of Technical Systems Evolution) that help engineers predict the most likely improvements that can be made to a given product. The most important of these laws involves the ideality of a system.

Substance-field Analysis

One more technique that is frequently used by inventors involves the analysis of substances, fields and other resources that are currently not being used and that can be found within the system or nearby. TRIZ uses non-standard definitions for substances and fields. Altshuller developed methods to analyze resources; several of his invention principles involve the use of different substances and fields that help resolve contradictions and increase ideality of a technical system. For instance, videotext systems used television signals to transfer data, by taking advantage of the small time segments between TV frames in the signals.

SuField analysis produces a structural model of the initial technological system, exposes its characteristics, and with the help of special laws, transforms the model of the problem. Through this transformation the structure of the solution that eliminates the shortcomings of the initial problem is revealed. SuField analysis is a special language of formulas with which it is possible to easily describe any technological system in terms of a specific (structural) model. A model produced in this manner is transformed according to special laws and regularities, thereby revealing the structural solution of the problem.

ARIZ - Algorithm of Inventive Problems Solving

ARIZ (Russian acronym of алгоритм решения изобретательских задач - АРИЗ) (algorithm of inventive problems solving) is a list of about 85 step-by-step procedures to solve complicated invention problems, where other tools of TRIZ alone (Sufield analysis, 40 inventive principles, etc.) are not sufficient.

Starting with an updated matrix of contradictions, semantic analysis, subcategories of inventive principles and lists of scientific effects, some new interactive applications are other attempts to simplify the problem formulation phase and the transition from a generic problem to a whole set of specific solutions.

Use of TRIZ on Management Problems

Although TRIZ was developed from the analysis of technical systems, it has been used widely as a method for understanding and solving complex management problems. Examples include finding additional cost savings for the legal department of a local government body: the inventive solution generated was to generate additional revenue [insert reference to cost-cutting in local government case study]. The results of the TRIZ work are expected to generate £1.7 m in profit in the first 5 years. TRIZ was also used to estimate the cost of nuclear decommissioning of buildings.

Use of TRIZ Methods in Industry

Case studies on the use of TRIZ are difficult to acquire as many companies believe TRIZ gives them a competitive advantage and are reluctant to publicise their adoption of the method. However some examples are available: Samsung is the most famous success story, and has invested heavily in embedding TRIZ use throughout the company, right up to and including the CEO; "In 2003 TRIZ led to 50 new patents for Samsung and in 2004 one project alone, a DVD pick-up innovation, saved Samsung over $100 million. TRIZ is now an obligatory skill set if you want to advance within Samsung."

Rolls-Royce, BAE Systems and GE are all documented users of TRIZ TRIZ is a Whizz article; Mars has documented how applying TRIZ led to a new patent for chocolate packaging.

TRIZ has also been used successfully by Leafield Engineering, Smart Stabilizer Systems and Buro Happold to solve problems and generate new patents.

Various promoters of TRIZ reported that car companies Rolls-Royce, Ford, and Daimler-Chrysler, Johnson & Johnson, aeronautics companies Boeing, NASA, technology companies Hewlett Packard, Motorola, General Electric, Xerox, IBM, LG, Samsung, Intel, Procter and Gamble, Expedia and Kodak have used TRIZ methods in some projects.

European TRIZ Association

The European TRIZ Association (ETRIA) is an association based in Germany, founded in 2000. ETRIA considers itself an open community to unite the efforts, suggest opportunities for global standardization, conduct further research and development, and provide mechanisms for the exchange of information and knowledge on TRIZ and TRIZ-based innovation technologies.

ETRIA is developing a web-based collaborative environment targeted the creation of links between any and all institutions concerned with conceptual questions pertaining to the creation, organization, and efficient processing of innovation knowledge and innovation technologies.

TRIZ is considered as a cross-disciplinary, generic methodology, but it has not previously been presented in terms of logic or any other formal knowledge representation. Most of the concepts introduced in TRIZ are fuzzy, and most of the techniques are still heuristic and only partially formalized. For further development and conceptual re-organization of the TRIZ knowledge base, ETRIA involves and collaborates with TRIZ experts and professionals from the domains of logic, organization science, informatics and linguistics.

Goals

ETRIA has the following goals:

- Research and development of innovation knowledge by integrating conceptual approaches to classification developed by artificial intelligence (AI) and knowledge management communities;
- International observation, analysis, evaluation and reporting of progress in these directions;
- Promotion and exchange of information and experience between scientists and practitioners in TRIZ, universities and other educational organizations;
- Development of TRIZ through contributions from dedicated experts and specialists in particular areas of expertise.

Modifications and derivatives

1. SIT (systematic inventive thinking)
2. USIT (unified structured inventive thinking)
3. Trizics (Methodology for the systematic application of TRIZ)

Social Entrepreneurship

Social entrepreneurship is the use of the techniques by start up companies and other entrepreneurs to develop, fund and implement solutions to social, cultural, or environmental issues. This concept may be applied to a variety of organizations with different sizes, aims, and beliefs. For-profit entrepreneurs typically measure performance using business metrics like profit, revenues and increases in stock prices, but social entrepreneurs are either non-profits or blend for-profit goals with generating a positive "return to society". and therefore must use different metrics. Social entrepreneurship typically attempts to further broad social, cultural, and environmental goals often associated with the voluntary sector in areas such as poverty alleviation, health care and community development.

Student organizers from the Green Club at Newcomb College Institute have formed a social entrepreneurship organization in 2010 that aimed to encourage people to reduce waste and live in a more environmentally conscious way.

At times, profit-making social enterprises may be established to support the social or cultural goals of the organization but not as an end in itself. For example, an organization that aims to provide housing and employment to the homeless may operate a restaurant, both to raise money and to provide employment for the homeless.

In the 2010s, social entrepreneurship is facilitated by the use of the Internet, particularly social networking and social media websites. These websites enable social entrepreneurs to reach a large number of people who are not geographically close yet who share the same goals and encourage them to collaborate online, learn about the issues, disseminate information about the group's events and activities, and raise funds through crowdfunding.

Modern Definition

In the 2000s, scholars and practitioners have debated which individuals or organizations can be considered to be social entrepreneurs. Thus far, there has been no firm consensus on the definition of social entrepreneurship, as so many different fields, disciplines and organization types are associated with social entrepreneurship, ranging from for-profit businesses to hybrid models combining charitable work with business activities, to non-profit charities, voluntary sector organizations and non-governmental organizations. Philanthropists, social activists, environmentalists, and other socially-oriented practitioners are often referred to as social entrepreneurs. Social entrepreneurs can include a range of career types and professional backgrounds, ranging from social work and community development to entrepreneurship and environmental science. For this reason, it is difficult to determine who is a social entrepreneur. David Bornstein has even used the term "social innovator" interchangeably with social entrepreneur, due to the creative, non-traditional strategies that many social entrepreneurs use. For a clearer definition of what social entrepreneurship entails, it is necessary to set the function of social entrepreneurship apart from other voluntary sector and charity-oriented activities and identify the boundaries within which social entrepreneurs operate. Some scholars have advocated restricting the term to founders of organizations that primarily rely on earned income (meaning income earned directly from paying consumers), rather than income from donations or grants. Others have extended this to include contracted work for public authorities, while still others include grants and donations.

Grameen Bank founder and Nobel Peace Prize winner Muhammad Yunus (left) with two young social entrepreneurs (to his right).

Social entrepreneurship in modern society offers an altruistic form of entrepreneurship that focuses on the benefits that society may reap. Simply put, entrepreneurship becomes a social endeavor when it transforms social capital in a way that affects society positively. It is viewed as advantageous because the success of social entrepreneurship depends on many factors related to social impact that traditional corporate businesses do not prioritize. Social entrepreneurs recognize immediate social problems, but also seek to understand the broader context of an issue that crosses disciplines, fields, and theories. Gaining a larger understanding of how an issue relates to society allows social entrepreneurs to develop innovative solutions and mobilize available resources to affect the greater global society. Unlike traditional corporate businesses, social entrepreneurship ventures focus on maximizing gains in social satisfaction, rather than maximizing profit gains. Both private and public agencies worldwide have had billion-dollar initiatives to empower deprived communities and individuals. Such support from organizations in society, such as government-aid agencies or private firms, may catalyze innovative ideas to reach a larger audience.

Prominent individuals associated with social entrepreneurship include Pakistani Akhter Hameed Khan and Bangladeshi Muhammad Yunus. Yunus was the founder of Grameen Bank, which pioneered the concept of microcredit for supporting innovators in multiple developing countries in Asia, Africa, and Latin America. He received a Nobel Peace Prize for his efforts. Others, such as former Indianapolis mayor Stephen Goldsmith addressed social efforts on a local level by using the private sector to provide city services.

History

Social entrepreneurship is distinct from the concept of entrepreneurship, yet still shares several similarities with its business cousin. Jean-Baptiste Say (1767–1832), a French economist, defined an entrepreneur as a person who "undertakes" an idea and shifts perspectives in a way that it alters the effect that an idea has on society. An entrepreneur is further defined by Say as someone who "shifts economic resources out of an area of lower and into an area of higher productivity and greater yield." The difference between "entrepreneurship" and "social entrepreneurship", however, stems from the purpose of a creation. Social entrepreneurs seek to transform societies at large, rather than transforming their profit margin, as classic entrepreneurs typically seek to do. Social entrepreneurs use a variety of resources to bring societies into a better state of well-being.

The concept of "social entrepreneurship" is not a novel idea, but in the 2000s, it has become more popular among society and academic research, notably after the publication of "The Rise of the Social Entrepreneur" by Charles Leadbeater. Many activities related to community development and higher social purpose fall within the modern definition of social entrepreneurship. Despite the established definition nowadays, social entrepreneurship remains a difficult concept to define, since it may be manifested in multiple forms. A broad definition of the concept allows interdisciplinary research efforts to understand and challenge the notions behind social entrepreneurship. No matter in which sector of society certain organizations are (i.e. corporations or unincorporated associations, societies, associations or cooperatives), social entrepreneurship focuses on the social impact that an endeavor aims at. Whether social entrepreneurship is altruistic or not is less important than the effect it has on society.

The terms *social entrepreneur* and *social entrepreneurship* were used first in the literature in 1953 by H. Bowen on his book "Social Responsibilities of the Businessman". The terms came into widespread use in the 1980s and 1990s, promoted by Bill Drayton, Charles Leadbeater, and others. From the 1950s to the 1990s, the politician Michael Young was a leading promoter of social entrepreneurship and in the 1980s, he was described by Professor Daniel Bell at Harvard University as the "world's most successful entrepreneur of social enterprises". Young created more than sixty new organizations worldwide, including the School for Social Entrepreneurs (SSE) which exists in the UK, Australia, and Canada and which supports individuals to realize their potential and to establish, scale, and sustain, social enterprises and social businesses. Another notable British social entrepreneur is Andrew Mawson OBE, who was given a peerage in 2007 because of his urban regeneration work including the Bromley by Bow Centre in East London. Although the terms are relatively new, social entrepreneurs and social entrepreneurship may be found throughout history. A list of a few noteworthy people whose work exemplifies the modern definition of "social entrepreneurship" includes Florence Nightingale, founder of the first nursing school and developer of modern nursing practices; Robert Owen, founder of the cooperative movement; and Vinoba Bhave, founder of India's Land Gift Movement. During the nineteenth and twentieth centuries some of the most successful social entrepreneurs straddled the civic, governmental, and business worlds. These pioneers promoted new ideas that were taken up by mainstream public services in welfare, schools, and health care.

2000s

Major Organizations

A panel discusses social entrepreneurship in the health care sector in 2015.

Groups focused on social entrepreneurship may be divided into several categories: community-based enterprises, socially responsible enterprises, social services industry professionals, and socio-economic enterprises. Community-based enterprises are based on the social ventures aimed at and involving an entire community. These enterprises build on the community's culture and capital (e.g., volunteer resources, financing, in-kind donations, etc.) to empower the enterprise and the community. Socially responsible enterprises focus on creating sustainable development through their initiatives that focus mostly on creating societal gains for the community. Social service industry professionals such as social workers and public health nurses work in social services, either for a government or a non-profit organization. They aim to expand social capital for individuals, communities, and organizations. Socio-economic enterprises include corporations that balance earning profits and with nonprofit goals, such as seeking social change for communities. Some social entrepreneurship organizations are not enterprises in a business sense; instead, they may be charities, non-profit organizations or voluntary sector organizations.

In addition, there are support organizations dedicated to empowering social entrepreneurs, connecting them with mentors, strengthening their enterprise models, and preparing them for capital investments. These incubators and accelerator organizations provide office and meeting space (often free), mentoring and coaching for social enterprise founders and leaders to help them develop their enterprises by improving the effectiveness of their business model, marketing, and strategy. Some accelerator organizations help social entrepreneur leaders to scale up their organization, either by taking it from a local scale to a national scale or from a national scale to a global scale. Some entrepreneurship support organizations also provide mentoring and coaching to social entrepreneurs.

One well-known social entrepreneur is Muhammad Yunus, who founded the Grameen Bank in 1976. He is known as the "father of microcredit," and established the microfinance movement, which aims to help millions of people rural communities to access small loans. For his work, he was awarded a Nobel Peace Prize in 2006. The work that Yunus did through Grameen Bank has been described as a major influence on later social entrepreneurs. Larger countries in Europe and South America have tended to work more closely with public organizations at both the national and local level.

Types

In *The Power of Unreasonable People*, John Elkington and Pamela Hartigan describe social entrepreneurs' business structures as falling under three different models, applicable in different situations and economic climates:

1. *Leveraged non-profit*: This business model leverages financial and other resources in an innovative way to respond to social needs.
2. *Hybrid non-profit*: This organizational structure can take a variety of forms, but is distinctive because the hybrid non-profit is willing to use profit from some activities to sustain its other operations which have a social or community purpose. Hybrid non-profits are often created to deal with government failures or market failures, as they generate revenue to sustain the operation without requiring loans, grants, and other forms of traditional funding.
3. *Social business venture*: These models are set up as businesses that are designed to create change through social means. Social business ventures evolved through a lack of funding. Social entrepreneurs in this situation were forced to become for-profit ventures, because loans and equity financing are hard to get for social businesses.

There are also a broader range of hybrid profit models, where a conventional business invests some portion of its profits on socially, culturally or environmentally beneficial activities. The term "Philanthropreneurship" has been applied to this type of activity. Corporate employees can also engage in social entrepreneurship, which may or may not be officially sanctioned by the company. This has been described as corporate social entrepreneurship.

One private foundation has staked the ground of more precise lexicon following the Newman's Own model having coined the phrase "Commercial Philanthropy" where commercial businesses are held and operated with all net proceeds going to serve social service needs.

International Presence

Organizations such as Ashoka, the Skoll Foundation, the Omidyar Network, the Schwab Foundation for Social Entrepreneurship, Tasamy, Athgo, New Profit Inc., National Social Entrepreneurship Forum, Echoing Green, and the Global Social Benefit Institute among others, focus on highlighting hidden change-makers and providing resources to advance their initiatives. The North American organizations tend to have a strongly individualistic stance focused on a handful of exceptional leaders. For example, The Skoll Foundation, created by eBay's first president, Jeff Skoll, makes capacity-building "mezzanine level" grants to social entrepreneurial organizations that already have reached a certain level of effectiveness.

Role of Technology

The Internet, social networking websites and social media have been pivotal resources for the success and collaboration of many social entrepreneurs. In the 2000s, the Internet has become especially useful in disseminating information to a wide range of like-minded supporters in short amounts of time, even if these individuals are geographically dispersed. In addition, the Internet allows for the pooling of design resources using open source principles. Using wiki models or crowdsourcing approaches, for example, a social entrepreneur organization can get hundreds of people from across a country (or from multiple countries) to collaborate on joint online projects (e.g., developing a business plan or a marketing strategy for a social entrepreneurship venture). These websites help social entrepreneurs to disseminate their ideas to broader audiences, help with the formation and maintenance of networks of like-minded people and help to link up potential investors, donors or volunteers with the organization. This enables social entrepreneurs to achieve their goals with little or no start-up capital and little or no "bricks and mortar" facilities (e.g., rented office space). For example, the rise of open-source appropriate technology as a sustainable development paradigm enables people all over the world to collaborate on solving local problems, just as open source software development leverages collaboration from software experts from around the world.

Public Opinion

Controversy

Many initiatives carried out with social entrepreneurs while innovative, have had problems becoming sustainable and effective initiatives that ultimately were able to branch out and reach the larger society as a whole (versus a small community or group of people). Studies of the qualities of social entrepreneurs have shown that very few individuals possess the talent and skills of entrepreneurs while at the same time having a primarily socially motivated outlook. Thus, compromises in social initiatives were developed, which often did not reach large audiences or help larger communities. Since the concept of social entrepreneurship has been popularized in the 2000s, some advocates suggest that there needs to be some standardization of the process in scaling up social endeavors to increase the impact of these ventures across the globe.

Policymakers around the globe may need to learn more about social initiatives, to increase the sustainability, effectiveness, and efficiency of these projects. Involvement and collaboration between private corporations and government agencies allows for increased support for carrying out social entrepreneurship initiatives, increased accountability on both ends, and increased connections with communities, individuals, or agencies in need. For example, private organizations or nonprofit organizations have tackled unemployment issues in communities. One challenge is that in some cases, social entrepreneurs may only propose short-term solutions, or that they are unable to scale up their virtual, online organization to a larger degree to maximize the number of people who are helped. Government program are able to tackle large issues; however, there is often little collaboration between governments and social entrepreneurs, which may have limited the effectiveness of social entrepreneurship. This lack of inter-sectoral collaboration may need to stagnation, if the motives and goals of social enterprises and of those in policy-making and programs are not aligned. Those in policy-making and the development of delivery of government programs tend to have different priorities than social entrepreneurs, resulting in slow growth and expansion of social initiatives.

Since social entrepreneurship has only started to gain momentum in the 2000s, current social entrepreneurs are encouraging social advocates and activists to develop into innovative social entrepreneurs. Increasing the scope and scale of social entrepreneurship may increase the likelihood of an efficient, sustainable, and effective initiative; although it may also render social entrepreneurship more challenging. Increased participation draws more attention to social entrepreneurship ventures from policymakers and privately owned corporations. The increased involvement from corporations and governments may help to strengthen social entrepreneurship, as it may lead to policy changes and to the development of training programs and leadership development programs for social entrepreneurs. Simultaneously, research shows that as social entrepreneurs attempt to widen their impact and scale their efforts, external institutions will have a key role to play in their success.

References

- Bennett, John L.; Bush, Mary Wayne (2013). Coaching for change. New York: Routledge. ISBN 9780415897815. OCLC 758394652.
- Fine, Alan; Merrill, Rebecca R (2010). You already know how to be great: a simple way to remove interference and unlock your greatest potential. New York: Portfolio Penguin. ISBN 9781591843559. OCLC 555648015.
- Landsberg, Max (2003) [1997]. The tao of coaching: boost your effectiveness at work by inspiring and developing those around you. London: Profile Books. ISBN 9781861976505. OCLC 223636717.
- Oettingen, Gabriele (2014). "The magic of WOOP". Rethinking positive thinking: inside the new science of motivation. New York: Current. pp. 118–140. ISBN 9781591846871. OCLC 881498635.
- Parsloe, Eric; Wray, Monika Jamieson (2000). Coaching and mentoring: practical methods to improve learning. London: Kogan Page. ISBN 9780749431181. OCLC 44058686.
- Scales, Peter (2008). Teaching in the lifelong learning sector. Maidenhead; New York: McGraw-Hill/Open University Press. ISBN 9780335222391. OCLC 179805428.
- Stamatis, D. H. (2001). Six sigma and beyond: foundations of excellent performance. 1. Boca Raton, FL: St. Lucie Press. ISBN 1574443119. OCLC 47100858.
- Goldsmith, Stephen (March 2010). The Power of Social Innovation: How Civic Entrepreneurs Ignite Community Networks for Good. Jossey-Bass. ISBN 978-0-470-57684-7.
- Hemingway, Christine A. Corporate Social Entrepreneurship: Integrity Within. Cambridge: Cambridge University Press, 2013. pg.135 ISBN 978-1-107-44719-6.
- "23 Creativity Killers That Will Drown Any Brainstorming Session". BoostCompanies. 2016-05-31. Retrieved 2016-10-24.

6

Technological Change: An Innovative Aspect

Technological change is the progress made in the production and consumption of technology. Some of the topics elucidated in this text are technological revolution, technological transitions, technological innovation system and technology scouting. This section helps the reader in developing an in-depth understanding of technological changes.

Technological Change

Technological change, technological development, technological achievement, or technological progress is the overall process of invention, innovation and diffusion of technology or processes. In essence technological change is the invention of technologies (including processes) and their commercialization via research and development (producing emerging technologies), the continual improvement of technologies (in which they often become less expensive), and the diffusion of technologies throughout industry or society (which sometimes involves disruption and convergence). In short, technological change is based on both better and more technology.

Modeling Technological Change

In its earlier days, technological change was illustrated with the 'Linear Model of Innovation', which has now been largely discarded to be replaced with a model of technological change that involves innovation at all stages of research, development, diffusion and use. When spoken about "modeling technological change" often the process of innovation is meant. This process of continuous improvement is often modeled as a curve depicting decreasing costs over time (for instance fuel cell which have become cheaper every year).

- TC is often modelled using a learning curve, ex.: Ct=Co * Xt^-b
- TC itself is often included in other models (for instance climate change models) and was often taken as an exogenous factor. These days TC is more often included as an endogenous factor. This means that it is taken as something you can influence. It is generally accepted that policy can influence the speed and direction of TC (for instance more towards clean technologies). This is referred to as Induced Technological Change.

Invention

The creation of something new, or a "breakthrough" technology. For example, a personal computer.

Diffusion

The spread of a technology through a society or industry. The diffusion of a technology generally follows an S-shaped curve as early versions of technology are rather unsuccessful, followed by a period of successful innovation with high levels of adoption, and finally a dropping off in adoption as a technology reaches its maximum potential in a market. In the case of a personal computer, it has made way beyond homes and into business settings, such as office workstations and server machines to host websites.

Technological Change as a Social Process

Underpinning the idea of *technological change as a social process* is general agreement on the importance of social context and communication. According to this model, technological change is seen as a social process involving producers and adopters and others (such as government) who are profoundly affected by cultural setting, political institutions and marketing strategies.

In free market economies, the maximization of profits is a powerful driver of technological change. Generally, only those technologies are developed and reach the market that promise to maximize profits for the owners of incoming producing capital. Any technologies that fail to meet this criterion even though they may satisfy very important societal needs, are not developed. Therefore, technological change is a social process strongly biased by the financial interests of capital. There are currently no well established democratic processes, such as voting on the social or environmental desirability of a new technology prior to development and marketing, that would allow average citizens to direct the course of technological change.

Elements of Diffusion

Emphasis has been on four key elements of the technological change process: (1) an innovative technology (2) communicated through certain channels (3) to members of a social system (4) who adopt it over a period of time. These elements are derived from Everett M. Rogers Diffusion of innovations theory using a communications-type approach.

Innovation

Rogers proposes that there are five main attributes of innovative technologies which influence acceptance, which he calls the ACCTO criteria. These are relative Advantage, Compatibility, Complexity, Trialability, and Observability. *Relative advantage* may be economic or non-economic, and is the degree to which an innovation is seen as superior to prior innovations fulfilling the same needs. It is positively related to acceptance (i.e., the higher the relative advantage, the higher the adoption level, and vice versa). *Compatibility* is the degree to which an innovation appears consistent with existing values, past experiences, habits and needs to the potential adopter; a low level of compatibility will slow acceptance. *Complexity* is the degree to which an innovation appears difficult to understand and use; the more complex an innovation, the slower its acceptance. *Trialability* is the perceived degree to which an innovation may be tried on a limited basis, and is positively related to acceptance. Trialability can accelerate acceptance because small-scale testing reduces risk. *Observability* is the perceived degree to which results of innovating are visible to others and is positively related to acceptance.

Communication Channels

Communication channels are the means by which a source conveys a message to a receiver. Information may be exchanged through two fundamentally different, yet complementary, channels of communication. Awareness is more often obtained through the *mass media*, while uncertainty reduction that leads to acceptance mostly results from *face-to-face communication*.

Social System

The social system provides a medium through which and boundaries within which, innovation is adopted. The structure of the social system affects technological change in several ways. Social norms, opinion leaders, change agents, government and the consequences of innovations are all involved. Also involved are cultural setting, nature of political institutions, laws, policies and administrative structures.

Time

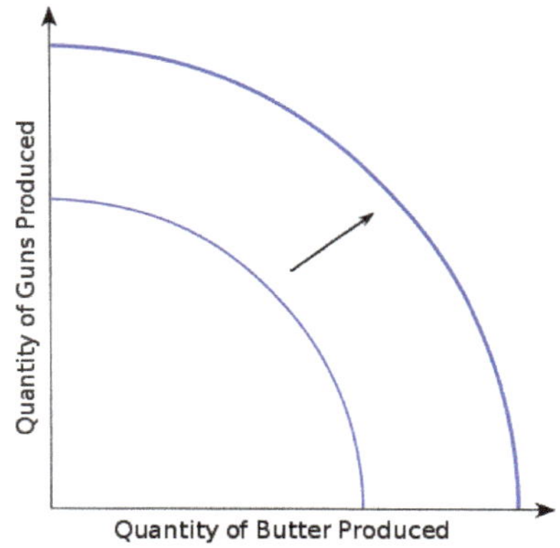

Technological change can cause the production-possibility frontier to shift outward, allowing economic growth.

Time enters into the acceptance process in many ways. The time dimension relates to the innovativeness of an individual or other adopter, which is the relative earlyness or lateness with which an innovation is adopted.

Economics

In economics, technological change is a change in the set of feasible production possibilities.

A technological innovation is Hicks neutral, following John Hicks (1932), if a change in technology does not change the ratio of capital's marginal product to labour's marginal product for a given capital-to-labour ratio. A technological innovation is Harrod neutral (following Roy Harrod) if the technology is labour-augmenting (i.e. helps labor); it is Solow neutral if the technology is capital-augmenting (i.e. helps capital).

Technological Revolution

Technological revolution is, in general, a relatively short period in history when one technology (or better a set of technologies) is replaced by another technology (or by the set of technologies). As Nick Bostrom wrote: *"We might define a technological revolution as a dramatic change brought about relatively quickly by the introduction of some new technology."* It is an era of an accelerated technological progress characterized not only by new innovations but also their application and diffusion.

Axe made of iron, dating from Swedish Iron Age, found at Gotland, Sweden. The iron - as a new material - initiated a dramatic revolution in technology, economy, society, warfare and politics.

A difference between technological revolution and technological change is not clearly defined. The technological change we could see as an introduction of an individual (single) new technology, while the technological revolution as a period in which *more* new technologies are adopted at the almost same time. These new technologies or technological changes are usually interconnected - as 3rd Kranzberg's law of technology says: *"Technology comes in packages, big and small."*

Description : By Joseph C. Galisim

A new technological revolution should increase a productivity of work, efficiency, etc. It may involve not only material changes but also changes in management, learning, social interactions, financing, methods of research etc. It is not limited strictly to technical aspects. Technological revolution so rewrites the material conditions of human existence and also reshape culture, society and even human nature. It can play a role of a trigger of a chain of various and unpredictable changes.

The Spinning Jenny and Spinning Mule (shown) greatly increased the productivity of thread manufacturing compared to the spinning wheel.

A Watt steam engine. The steam engine, fuelled primarily by coal, propelled the Industrial Revolution in Great Britain and the world.

IBM Personal Computer XT in 1988 - The PC was an invention that dramatically changed not only a professional but also personal life.

"What distinguishes a technological revolution from a random collection of technology systems and justifies conceptualizing it as a revolution are two basic features:

1. The strong interconnectedness and interdependence of the participating systems in their technologies and markets.

2. The capacity to transform profoundly the rest of the economy (and eventually society)."

The consequences of a technological revolution are not exclusively positive but also not that negative :3 - for example, it can have negative environmental impact and cause a temporal unemployment (so called technological unemployment).

The concept of technological revolution is based on the idea (not unquestioned) that technological progress is not linear but undulatory. Technological revolution can be:

- sectoral (more technological changes in one sector, e.g. Green revolution, Commercial revolution)
- universal (interconnected radical changes in more sectors, the universal technological revolution can be seen as a complex of several parallel sectoral technological revolutions, e.g. Second industrial revolution, Renaissance technological revolution etc.)

The concept of universal technological revolutions is a key factor in the Neo-Schumpeterian theory of long economic waves/cycles (Carlota Perez, Tessaleno Devezas, Daniel Šmihula and others).

History

The most known example of technological revolution was the Industrial revolution in the 19th century, the Scientific-technical revolution about 1950 - 1960, the Neolithic revolution, the Digital revolution etc. The notion of "technological revolution" is frequently overused. Therefore, it is not easy to define which technological revolutions having occurred during world history were really crucial and influenced not only one segment of human activity but had a universal impact. One universal technological revolution should be composed from several sectoral technological revolutions (in science, industry, transport etc.).

We can identify several universal technological revolutions which occurred during the modern era in Western culture:

- 1. (1600–1740) Financial-agricultural revolution
- 2. (1780–1840) Industrial revolution
- 3. (1870–1920) Technical revolution (or Second Industrial Revolution)

- 4. (1940–1970) Scientific-technical revolution
- 5. (1985–present) Information and telecommunications revolution

Attempts to find comparable periods of well defined technological revolutions in the pre-modern era are highly speculative. Probably one of the most systematic attempts to suggest a timeline of technological revolutions in pre-modern Europe was done by Daniel Šmihula:

- A. (1900-1100 BC) Indo-European technological revolution
- B. (700- 200 BC) Celtic and Greek technological revolution
- C. (300- 700 AD) Germano-Slavic technological revolution
- D. (930-1200 AD) Medieval technological revolution
- E. (1340-1470 AD) Renaissance technological revolution

The potential Future Technological Revolution

After 2000 there became popular the idea that a sequence of technological revolutions is not over and in the forthcoming future we will witness the dawn of a new universal technological revolution. The main innovations should develop in the fields of nanotechnologies, alternative fuel and energy systems, biotechnologies, genetic engineering, new materials technologies etc.

Relation to "Technological Revolution" and "Technical Revolution"

Sometimes the notion of "Technological revolution" is used for the Second Industrial Revolution in the period about 1900. But in this case the designation "Technical revolution" would be more proper. When the notion of technical revolution is used in more general meaning it is almost identical with technological revolution but technological revolution requires material changes in used tools, machines, energy sources, production processes. Technical revolution can be restricted to changes in management, organisation and so called non-material technologies (e.g. a progress in mathematics or accounting).

List of intellectual, Philosophical and Technological Revolutions (Sectoral or Universal)

- The Upper Paleolithic Revolution: The emergence of "high culture", new technologies and regionally distinct cultures (50,000 - 40,000 years ago).
- The Neolithic Revolution (perhaps 13000 years ago), which formed the basis for human civilization to develop.

- The Renaissance technological revolution: The set of inventions during the Renaissance period, roughly the 14th through the 16th century.
- The Commercial Revolution: A period of European economic expansion, colonialism, and mercantilism which lasted from approximately the 16th century until the early 18th century.
- The Price revolution: A series of economic events from the second half of the 15th century to the first half of the 17th, the price revolution refers most specifically to the high rate of inflation that characterized the period across Western Europe.
- The Scientific revolution: A fundamental transformation in scientific ideas around the 16th century.
- The British Agricultural Revolution (18th century), which spurred urbanisation and consequently helped launch the Industrial Revolution.
- The Industrial Revolution: The major shift of technological, socioeconomic and cultural conditions in the late 18th century and early 19th century that began in Britain and spread throughout the world.
- The Market Revolution: A drastic change in the manual labor system originating in the South of the United States (and soon moving to the North) and later spreading to the entire world (about 1800–1900).
- The Second Industrial Revolution (1871–1914).
- The Green Revolution (1945-1975): The use of industrial fertilizers and new crops greatly increased the world's agricultural output.
- The Digital Revolution: The sweeping changes brought about by computing and communication technology, starting from circa 1950 with the creation of the first general-purpose electronic computers.
- The Information revolution: The massive economic, social and technological changes resulting from the Digital revolution (after 1960?).

Technological Transitions

Technological innovations have occurred throughout history and rapidly increased over the modern age. New technologies are developed and co-exist with the old before supplanting them. Transport offers several examples; from sailing to steam ships to automobiles replacing horse-based transportation. Technological transitions (TT) describe how these technological innovations occur and are incorporated into society.

Alongside the technological developments TT considers wider societal changes such as "user practices, regulation, industrial networks (supply, production, distribution), infrastructure, and symbolic meaning or culture". For a technology to have use, it must be linked to social structures human agency and organisations to fulfil a specific need. Hughes refers to the 'seamless web' where physical artefacts, organisations, scientific communities, and social practices combine. A technological system includes technical and non-technical aspects, and it a major shift in the socio-technical configurations (involving at least one new technology) is when a technological transition occurs.

Origins

Work on technological transitions draws on a number of fields including history of science, technology studies, and evolutionary economics. The focus of evolutionary economics is on economic change, but as a driver of this technological change has been considered in the literature. Joseph Schumpeter, in his classic *Theory of Economic Development* placed the emphasis on non-economic forces as the driver for growth. The human actor, the entrepreneur is seen as the cause of economic development which occurs as a cyclical process. Schumpeter proposed that radical innovations were the catalyst for Kondratiev cycles.

Long Wave Theory

The Russian economist Kondratiev proposed that economic growth operated in boom and bust cycles of approximately 50 year periods. These cycles were characterised by periods of expansion, stagnation and recession. The period of expansion is associated with the introduction of a new technology, e.g. steam power or the microprocessor. At the time of publication, Kondratiev had considered that two cycles had occurred in the nineteenth century and third was beginning at the turn of the twentieth. Modern writers, such as Freeman and Perez outlined five cycles in the modern age:

- The Industrial Revolution (1770–1830)
- Victorian Prosperity: Age of steam and Rail (1830–1880)
- The Age of Steel (1880–1930)
- Oil, Mass Production and the Consumer Society (1930–1980)
- The Information Age (1980-?)

Freeman and Perez proposed that each cycle consists of pervasive technologies, their production and economic structures that support them. Termed 'techno-economic paradigms', they suggest that the shift from one paradigm to another is the result of emergent new technologies.

Following the recent economic crisis, authors such as Moody and Nogrady have suggested that a new cycle is emerging from the old, centred on the use of sustainable technologies in a resource depleted world.

Technological Paradigms, Trajectories and Regimes

Thomas Kuhn described how a paradigm shift is a wholesale shift in the basic understanding of a scientific theory. Examples in science include the change of thought from miasma to germ theory as a cause of disease. Building on this work, Giovanni Dosi developed the concept of 'technical paradigms' and 'technological trajectories'. In considering how engineers work, the technical paradigm is an outlook on the technological problem, a definition of what the problems and solutions are. It charts the idea of specific progress. By identifying the problems to be solved the paradigm exerts an influence on technological change. The pattern of problem solving activity and the direction of progress is the technological trajectory. In similar fashion, Nelson and Winter (,)defined the concept of the 'technological regime' which directs technological change through the beliefs of engineers of what problems to solve. The work of the actors and organisations is the result of organisational and cognitive routines which determines search behaviour. This places boundaries and also trajectories (direction) to those boundaries.

Multi-level Perspective on Technological Transitions

In analysing (historic) cases of technological transitions researchers from the systems in transition branch of transitions research have used a multi-level perspective (MLP) as a heuristic model to understand changes in socio-technical systems. () Innovation system approaches traditionally focus on the production side. A socio-technical approach combines the science and technology in devising a production, with the application of the technology in fulfilling a societal function. Linking the two domains are the distribution, infrastructure and markets of the product. This approach considers a transition to be multi-dimensional as technology is only one aspect.

The MLP proposes three analytical levels: the niche, regime and landscape.

Niche (Micro-level) Radical innovations occur at the niche level. These act as 'safe havens' for fledgling technologies to develop, largely free from market pressures which occur at the regime level. The US Military has acted as niche for major twentieth century technologies such as the aircraft, radio and the internet. More recently, California's Silicon Valley has provided an arena for ICT focused technologies to emerge. Some innovations will challenge the existing regime while others fail.

Regime (Meso-level) The socio-technical regime, as defined by Geels, includes a web of inter-linking actors across different social groups and communities following a set of rules. In effect, the established practices of a given system. Seven dimensions have been identified in the socio-technical regime: technology, user practices and application, the symbolic meaning of technology, infrastructure, policy and techno-scientific knowledge. Change does occur at the regime level but it is normally slow and incremental unlike the radical change at the niche level. The actors who constitute the existing regime are set to gain from perpetuating the incumbent technology at the expense of the new. This is known as 'lock-in'.

Landscape (Macro-level) Exogenous to the previous levels is the socio-technical landscape. A broad range of factors are contained here, such as economic pressures, cultural values, social trends, wars and environmental issues. Change occurs at an even slower rate than at the regime level.

A transition is said to happen when a regime shift has occurred. This is the result of the interplay between the three levels. Regimes are relatively inert and resistant to change being structured to incremental innovation following established trajectories. As such, transitions are difficult to achieve. The current regime is typically suffering internal issues. Pressure from the landscape level may cause 'cracks' or 'windows of opportunity' through which innovations at the niche level may initially co-exist with the established technology before achieving ascendency. Once the technology has fully embedded into society the transition is said to be completed.

Case Study

The MLP has been used in describing a range of historic transitions in socio-technical regimes for mobility, sanitation, food, lighting and so on. While early research focused on historical transitions, a second strand of research was more focused on transitions to sustainable technologies in key sectors such as transport, energy and housing.

Geels presented three historical transitions on system innovation relating to modes of transportation. The technological transition from sailing ships to steamships in the UK will be summarised and shown in the context of a wider system innovation.

Great Britain was the world's leading naval power in the nineteenth century, and led the way in the transition from sail to steam. At first, the introduction of steam technology co-existed with the current regime. Steam tugs assisted sail ships into port and hybrid steam / sail ships appeared. Landscape developments create the necessity for improvements in the technology. A demand for trans-Atlantic emigration was prompted by the Irish potato famine, European political instability and the lure of gold in California. The requirement for such arduous journeys had prompted a wealth of innovations at the niche level in steamship-development. From the late 1880s, as steamship technology improved and costs dropped, the new technology was widely diffused and a new regime established. The changes go beyond a technological transition as it involved new ship management and fleet management practices, new supporting infrastructures and new functionalities.

Transition Paths

The nature of transitions varies and the differing qualities result in multiple pathways occurring. Geels and Schot defined five transition paths:

- Reproduction: Ongoing change occurring in the regime level.
- Transformation: A socio-technical regime that changes without the emergence of a monopolising technology.
- Technological substitution: An incumbent technology is replaced by a radical innovation resulting in a new socio-technical regime. (E.g. the automobile replacing the horse as the primary means of land transport).
- De-alignment and Re-alignment: Weaknesses in the regime sees the advent of competing new technologies leading to a dominant model.
- Re-configuration: When multiple, interlinked technologies are replaced by a similarly linked alternative set.

Characteristics of Technological Transitions

Six characteristics of technological transitions have been identified.,

Transitions are co-evolutionary and multi-dimensional Technological developments occur intertwined with societal needs, wants and uses. A technology is adopted and diffused based on this interplay between innovation and societal requirements. Co-evolution has different aspects. As well as the co-evolution of technology and society, aspects between science, technology, users and culture have been considered.

Multi-actors are involved Scientific and engineering communities are central to the development of a technology, but a wide range of actors are involved in a transition. This can include organisations, policy-makers, government, NGOs, special interest groups and others.

Transitions occur at multiple levels As shown in the MLP transitions occur through the interplay of processes at different levels.

Transitions are a long-term process Complete system-change takes time and can be decades in the making. Case studies show them to be between 40 and 90 years.

Transitions are radical For a true transition to occur the technology has to be a radical innovation.

Change is Non-linear The rate of change will vary over time. For example, the pace of change may be slow at the gestation period (at the niche level) but much more rapid when a breakthrough is occurring.

Diffusion: Transition Phases

Diffusion of an innovation is the concept of how it is picked up by society, at what rate and why. Everett (1962).The diffusion of a technological innovation into society can be considered in distinct phases. Pre-development is the gestation period where the new technology has yet to make an impact. Take-off is when the process of a system shift is beginning. A breakthrough is occurring when fundamental changes are occurring in existing structures through the interplay of economic, social and cultural forces. Once the rate of change has decreased and a new balance is achieved, stabilization is said to have occurred. A full transition involves an overhaul of existing rules and change of beliefs which takes time, typically spanning at least a generation. This process can be speeded-up through seismic, unforeseen events such as war or economic strife.

Geels proposed a similar four phased approach which draws on the multi-level perspective (MLP) developed by Dutch scholars. Phases one sees the emergence of a novelty, born from the existing regime. Development then occurs in the niche level at phase two. As before, breakthrough then occurs at phase three. In the parlance of the MLP the new technology, having been developed at the niche level, is in competition with the established regime. To breakthrough and achieve wide diffusion, external factors – 'windows of opportunity' are required.

Windows of Opportunity

A number of possible circumstances can act as windows of opportunity for the diffusion of new technologies:

- Internal technical problems in the existing regime. Those that cannot be solved by refinement of existing technologies act as a driver for the new.
- Problems external to the system. Such 'problems' are often determined by pressure groups and require wider societal or political backing. An example is environmental concerns.
- Changing user preferences. Opportunities are presented if existing technologies cannot meet user needs.
- Strategic advantage. Competition with rivals may necessitate innovation
- Complimentary technology. The availability of which may enable a breakthrough

Alongside external influences, internal drivers catalyse diffusion. These include economic factors such as the price performance ration. Socio-technical perspectives focus on the links between disparate social and technological elements.

Societal Relevance

The study of technological transitions has an impact beyond academic interest. The transitions referred to in the literature may relate to historic processes, such as the transportation transitions studied by Geels, but system changes are required to achieve a safe transition to a low carbon-economy. Current structural problems are apparent in a range of sectors. Dependency on oil is problematic in the energy sector due to availability, access and contribution to greenhouse gas (GHG) emissions. Transportation is a major user of energy causing significant emission of GHGs. Food production will need to keep pace with an ever-growing world population while overcoming challenges presented by global warming and transportation issues. Incremental change has provided some improvements but a more radical transition is required to achieve a more sustainable future.

Developed from the work on technological transitions is the field of transition management. Within this is an attempt to shape the direction of change complex socio-technical systems to more sustainable patterns. Whereas work on technological transitions is largely based on historic processes, proponents of transition management seek to actively steer transitions in progress.

Criticisms

Genus and Coles outlined a number of criticisms against the analysis of technological transitions, in particular when using the MLP. Empirical research on technological transitions occurring now has been limited, with the focus on historic transitions. Depending on the perspective on transition case studies they could be presented as having occurred on a different transition path to what was shown. For example, the bicycle could be considered an intermediate transport technology between the horse and the car. Judged from shorter different time-frame this could appear a transition in its own right. Determining the nature of a transition is problematic; when it started and ended, or whether one occurred in the sense of a radical innovation displacing an existing socio-technical regime. The perception of time casts doubt on whether a transition has occurred. If viewed over a long enough period even inert regimes may demonstrate radical change in the end. The MLP has also been criticised by scholars studying sustainability transitions using Social Practice Theories.

Technological Innovation System

The technological innovation system is a concept developed within the scientific field of innovation studies which serves to explain the nature and rate of technological change. A Technological Innovation System can be defined as 'a dynamic network of agents interacting in a specific economic/industrial area under a particular institutional infrastructure and involved in the generation, diffusion, and utilization of technology'.

The approach may be applied to at least three levels of analysis: to a technology in the sense of a knowledge field, to a product or an artifact, or to a set of related products and artifacts aimed at satisfying a particular (societal) function'. With respect to the latter, the approach has especially proven itself in explaining why and how sustainable (energy) technologies have developed and diffused into a society, or have failed to do so.

Background

The concept of a technological innovation system was introduced as part of a wider theoretical school, called the innovation system approach. The central idea behind this approach is that determinants of technological change are not (only) to be found in individual firms or in research institutes, but (also) in a broad societal structure in which firms, as well as knowledge institutes, are embedded. Since the 1980s, innovation system studies have pointed out the influence of societal structures on technological change, and indirectly on long-term economic growth, within nations, sectors or technological fields.

The purpose of analyzing a Technological Innovation System is to analyze and evaluate the development of a particular technological field in terms of the structures and processes that support or hamper it. Besides its particular focus, there are two, more analytical, features that set the Technological Innovation System approach apart from other innovation system approaches.

Firstly, the Technological Innovation System concept emphasizes that stimulating knowledge flows is not sufficient to induce technological change and economic performance. There is a need to exploit this knowledge in order to create new business opportunities. This stresses the importance of individuals as sources of innovation, something which is sometimes overseen in the, more macro-oriented, nationally or sectorally oriented innovation system approaches.

Secondly, the Technological Innovation System approach often focuses on system dynamics. The focus on entrepreneurial action has encouraged scholars to consider a Technological Innovation System as something to be built up over time. This was already put forward by Carlsson and Stankiewicz:

'[T]echnological Innovation Systems are defined in terms of knowledge/competence flows rather than flows of ordinary goods and services. They consist of dynamic knowledge and competence networks. In the presence of an entrepreneur and sufficient critical mass, such networks can be transformed into development blocks, i.e. synergistic clusters of firms and technologies within an industry or a group of industries.'

This means that a Technological Innovation System may be analyzed in terms of its system components and/or in terms of its dynamics. Both perspectives will be explained below.

Structures

The system components of a Technological Innovation System are called structures.

These represent the static aspect of the system, as they are relatively stable over time. Three basic categories are distinguished:

- Actors: Actors involve organizations contributing to a technology, as a developer or adopter, or indirectly as a regulator, financier, etc. It is the actors of a Technological Innovation System that, through choices and actions, actually generate, diffuse and utilize technologies. The potential variety of relevant actors is enormous, ranging from private actors to public actors, and from technology developers to technology adopters. The development of a Technological Innovation System will depend on the interrelations between all these actors. For example, entrepreneurs are unlikely to start investing in their businesses if governments are unwilling to support them financially. Visa-verse, governments have no clue where financial support is necessary if entrepreneurs do not provide them with the information and the arguments they need to legitimate policy support.

- Institutions: Institutional structures are at the core of the innovation system concept. It is common to consider institutions as 'the rules of the game in a society, or, more formally, (...) the humanly devised constraints that shape human interaction'. A distinction can be made between formal institutions and informal institutions, with formal institutions being the rules that are codified and enforced by some authority, and informal institutions being more tacit and organically shaped by the collective interaction of actors. Informal institutions can be normative or cognitive. The normative rules are social norms and values with moral significance, whereas cognitive rules can be regarded as collective mind frames, or social paradigms. Examples of formal institutions are government laws and policy decisions; firm directives or contracts also belong to this category. An example of a normative rule is the responsibility felt by a company to prevent or clean up waste. Examples of cognitive rules are search heuristics or problem-solving routines. They also involve dominant visions and expectations held by the actors.

- Technological factors: Technological structures consist of artefacts and the technological infrastructures in which they are integrated. They also involve the techno-economic workings of such artefacts, including costs, safety, reliability. These features are crucial for understanding the feedback mechanisms between technological change and institutional change. For example, if R&D subsidy schemes supporting technology development should result in improvements with regard to the safety and reliability of applications, this would pave the way for more elaborate support schemes, including practical demonstrations. These may, in turn, benefit technological improvements even more. It should, however, be noted here that the importance of technological features has often been neglected by scholars.

The structural factors are merely the elements that make up the system. In an actual system, these factors are all linked to each other. If they form dense configurations they are called networks. An example would be a coalition of firms jointly working on the application of a fuel cell, guided by a set of problem-solving routines and supported by a subsidy program. Likewise, industry associations, research communities, policy networks, user-supplier relations etc. are all examples of networks.

An analysis of structures typically yields insight into systemic features - complementarities and conflicts - that constitute drivers and barriers for technology diffusion at a certain moment or within a given period in time.

Dynamics

Structures involve elements that are relatively stable over time. Nevertheless, for many technologies, especially newly emerging ones, these structures are not yet (fully) in place. For this reason, mostly, the scholars have recently enriched the literature on Technological Innovation Systems with studies that focus on the build-up of structures over time. The central idea of this approach is to consider all activities that contribute to the development, diffusion, and use of innovations as system functions. These system functions are to be understood as types of activities that influence the build-up of a Technological Innovation System. Each system function may be 'fulfilled' in a variety of ways. The premise is that, in order to properly develop, the system should positively fulfil all system functions. Various 'lists' of system functions have been constructed. Authors like Bergek et al., Hekkert et al., Negro and Suurs give useful overviews. These lists show much overlap and differences reside mostly in the particular way of clustering activities. An example of such a list is provided below.

Note that it is also possible that activities negatively contribute to a system function. These negative contributions imply a (partial) breakdown of the system.

Seven System Functions

As an example, the seven system functions defined by Hekkert are explained here:

- F1. Entrepreneurial activities: The classic role of the entrepreneur is to translate knowledge into business opportunities, and eventually innovations. The entrepreneur does this by performing market-oriented experiments that establish change, both to the emerging technology and to the institutions that surround it. The Entrepreneurial Activities involve projects aimed to prove the usefulness of the emerging technology in a practical and/or commercial environment. Such projects typically take the form of experiments and demonstrations.

- F2. Knowledge development: The Knowledge Development function involves learning activities, mostly on the emerging technology, but also on markets, networks, users etc. There are various types of learning activities, the most important categories being learning-by-searching and learning-by-doing. The former concerns R&D activities in basic science, whereas the latter involves learning activities in a practical context, for example in the form of laboratory experiments or adoption trials.

- F3. Knowledge diffusion / knowledge exchange through networks: The characteristic organization structure of a Technological Innovation System is that of the network. The primary function of networks is to facilitate the exchange of knowledge between all the actors involved in it. Knowledge Diffusion activities involve partnerships between actors, for example technology developers, but also meetings like workshops and conferences. The important role of Knowledge Diffusion stems from Lundvall's notion of interactive learning as the raison-d'être of any innovation system. The innovation system approach stresses that innovation happens only where actors of different backgrounds interact. A special form of interactive learning is learning-by-using, which involves learning activities based on the experience of users of technological innovations, for example through user-producer interactions.

- F4. Guidance of the search: The Guidance of the Search function refers to activities that shape the needs, requirements and expectations of actors with respect to their support of the emerging technology. Guidance of the Search refers to individual choices related to the technology but it may also take the form of hard institutions, for example policy targets. It also refers to promises and expectations as expressed by various actors in the community. Guidance of the Search can be positive or negative. A positive Guidance of the Search means a convergence of positive signals - expectations, promises, policy directives - in a particular direction of technology development. If negative, there will be a digression, or, even worse, a rejection of development altogether. This convergence is important since, usually, various technological options exist within an emerging technological field, all of which require investments in order to develop further. Since resources are usually limited, it is important that specific foci are chosen. After all, without any focus there will be a dilution of resources, preventing all options from prospering. On the other hand, too much focus may result in the loss of variety. A healthy Technological Innovation System will strike a balance between creating and reducing variety.

- F5. Market formation: Emerging technologies cannot be expected to compete with incumbent technologies. In order to stimulate innovation, it is usually necessary to create artificial (niche) markets. The Market Formation function involves activities that contribute to the creation of a demand for the emerging technology, for example by financially supporting the use of the emerging technology, or by taxing the use of competing technologies. Market Formation is especially important in the field of sustainable energy technologies, since, in this case, there usually is a strong normative legitimation for the intervention in market dynamics.

- F6. Resource mobilization: Resource Mobilization refers to the allocation of financial, material and human capital. The access to such capital factors is necessary for all other developments. Typical activities involved in this system function are investments and subsidies. They can also involve the deployment of generic infrastructures such as educational systems, large R&D facilities or refueling infrastructures. In some cases, the mobilization of natural resources, such as biomass, oil or natural gas is important as well. The Resource Mobilization function represents a basic economic variable. Its importance is obvious: an emerging technology cannot be supported in any way if there are no financial or natural means, or if there are no actors present with the right skills and competences.
- F7. Support from advocacy coalitions: The rise of an emerging technology often leads to resistance from actors with interests in the incumbent energy system. In order for a Technological Innovation System to develop, other actors must counteract this inertia. This can be done by urging authorities to reorganize the institutional configuration of the system. The Support from Advocacy Coalitions function involves political lobbies and advice activities on behalf of interest groups. This system function may be regarded as a special form of Guidance of the Search. After all, lobbies and advices are pleas in favor of particular technologies. The essential feature which sets this category apart is that advocacy coalitions do not have the power, like for example governments, to change formal institutions directly. Instead, they employ the power of persuasion. The notion of the advocacy coalition is based on the work of Sabatier, who introduced the idea within the context of political science. The concept stresses the idea that structural change within a system is the outcome of competing interest groups, each representing a separate system of values and ideas. The outcome is determined by political power.

Cumulative Causation

Since Carlsson and Stankiewicz introduced the concept of a Technological Innovation System, an increasing number of scholars have started focusing on dynamics. A recurring theme within their studies has been the notion of cumulative causation, closely related to the idea of a virtuous circle or vicious circle, by Gunnar Myrdal.

In this context, cumulative causation is the phenomenon that the build-up of a Technological Innovation System accelerates due to system functions interacting and reinforcing each other over time. For example, the successful realization of a research project, contributing to Knowledge Development, may result in high expectations, contributing to Guidance of the Search, among policy makers, which may, subsequently, trigger the start-up of a subsidy program, contributing to Resource Mobilization, which induces even more research activities: Knowledge Development, Guidance of the Search, etc. System functions may also reinforce each other 'downwards'. In that case interactions result in conflicting developments or a vicious circle! Recently scholars have increasingly paid attention to the question of how cumulative causation may be established, often with a particular focus on the development of sustainable energy technologies.

Acquiring New Technologies and Capabilities

To improve competitiveness and retain sustainability, firms require new technologies and capabilities. In this age of rapid innovation and complexity, it is challenging for the firms to develop internally and remain competitive at the same time. Merger, acquisition and alliance are some of the ways to achieve this, but the primary driver is the desire to obtain valuable resources. Many acquisitions failed to achieve their objectives and resulted in poor performance because of improper implementation.

1. Improper documentation and changing implicit knowledge makes it difficult to share information during acquisition.

2. For acquired firm symbolic and cultural independence which is the base of technology and capabilities are more important than administrative independence.

3. Detailed knowledge exchange and integrations are difficult when the acquired firm is large and high performing.

4. Management of executives from acquired firm is critical in terms of promotions and pay incentives to utilize their talent and value their expertise.

5. Transfer of technologies and capabilities are most difficult task to manage because of complications of acquisition implementation. The risk of losing implicit knowledge is always associated with the fast pace acquisition.

Preservation of tacit knowledge, employees and literature are always delicate during and after acquisition. Strategic management of all these resources is a very important factor for a successful acquisition.

Increase in acquisitions in our global business environment has pushed us to evaluate the key stake holders of acquisition very carefully before implementation. It is imperative for the acquirer to understand this relationship and apply it to its advantage. Retention is only possible when resources are exchanged and managed without affecting their independence.

Technology Scouting

Technology scouting can be regarded as a method of technology forecasting or in the broader context also an element of corporate foresight. At the same time Technology Scouting also contributes to technology management by (1) identifying emerging technologies, (2) channel technology related information into an organization, and (3) in a corporate context supports the acquisition of technologies. It is a mechanism to practically implement and managing open innovation. In this context, technology scouting is the starting point of a long term and interactive matching process between external technologies and internal requirements of an existing organization for strategic purposes.

Definitions

Technology Scout

The technology scout is either an employee of the company or an external consultant. He or she may be assigned part-time or full-time to the scouting task. The desired characteristics of a technology scout are similar to the characteristics associated with the technological gatekeeper.

These characteristics include being a lateral thinker, knowledgeable in science and technology, respected inside the company, cross-disciplinary orientated, and imaginative personality.

Technology Scouting

Technology scouting is a systematic approach by companies whereby they assign part of their staff or employ external consultants to gather information in the field of science and technology and through which they facilitate or execute technology sourcing. Technology scouting is either directed at a specific technological area or undirected, identifying relevant developments in technological white spaces. Technology scouting relies on formal and informal information sources, including the personal networks of the scouts.

Kondratiev Wave

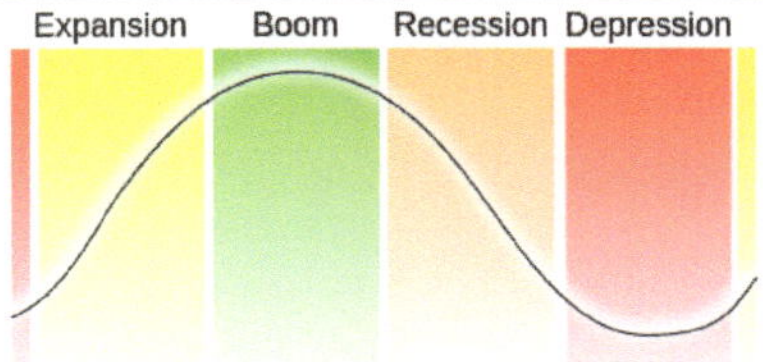

Proposed Economic Waves

Cycle/Wave Name	Period (years)
Kitchin cycle (inventory, e.g. pork cycle)	3–5
Juglar cycle (fixed investment)	7–11
Kuznets swing (infrastructural investment)	15–25
Kondratiev wave (technological basis)	45–60

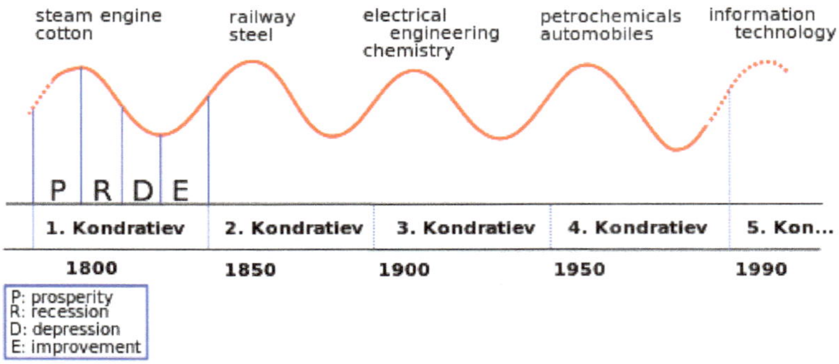

A rough schematic drawing showing growth cycles in the world economy over time according to the Kondratiev theory

In economics, Kondratiev waves (also called supercycles, great surges, long waves, K-waves or the long economic cycle) are supposed cycle-like phenomena in the modern world economy.

It is stated that the period of a wave ranges from forty to sixty years, the cycles consist of alternating intervals between high sectoral growth and intervals of relatively slow growth.

History of Concept

The Soviet economist Nikolai Kondratiev (also written Kondratieff) was the first to bring these observations to international attention in his book *The Major Economic Cycles* (1925) alongside other works written in the same decade. In 1939, Joseph Schumpeter suggested naming the cycles "Kondratieff waves" in his honor.

Two Dutch economists, Jacob van Gelderen and Salomon de Wolff, had previously argued for the existence of 50- to 60-year cycles in 1913.

Since the inception of the theory, various studies have expanded the range of possible cycles, finding longer or shorter cycles in the data. The Marxist scholar Ernest Mandel revived interest in long-wave theory with his 1964 essay predicting the end of the long boom after five years and in his Alfred Marshall lectures in 1979. However, in Mandel's theory, there are no long "cycles", only distinct epochs of faster and slower growth spanning 20–25 years.

The historian Eric Hobsbawm wrote of the theory: "That good predictions have proved possible on the basis of Kondratiev Long Waves—this is not very common in economics—has convinced many historians and even some economists that there is something in them, even if we don't know what."

Characteristics of the Cycle

Kondratiev identified three phases in the cycle: expansion, stagnation, and recession. More common today is the division into four periods with a turning point (collapse) between the first and second phases. Writing in the 1920s, Kondratiev proposed to apply the theory to the 19th century:

- 1790–1849 with a turning point in 1815.
- 1850–1896 with a turning point in 1873.
- Kondratiev supposed that, in 1896, a new cycle had started.

The long cycle supposedly affects all sectors of an economy. Kondratiev focused on prices and interest rates, seeing the ascendant phase as characterized by an increase in prices and low interest rates, while the other phase consists of a decrease in prices and high interest rates. Subsequent analysis concentrated on output.

Explanations of the Cycle

Technological Innovation Theory

According to the innovation theory, these waves arise from the bunching of basic innovations that launch technological revolutions that in turn create leading industrial or commercial sectors. Kondratiev's ideas were taken up by Joseph Schumpeter in the 1930s. The theory hypothesized the existence of very long-run macroeconomic and price cycles, originally estimated to last 50–54 years.

In recent decades there has been considerable progress in historical economics and the history of technology, and numerous investigations of the relationship between technological innovation and economic cycles. Some of the works involving long cycle research and technology include Mensch (1979), Tylecote (1991), The International Institute for Applied Systems Analysis (IIASA) (Marchetti, Ayres), Freeman and Louçã (2001), Andrey Korotayev and Carlota Perez.

Perez (2002) places the phases on a logistic or *S* curve, with the following labels: the beginning of a technological era as *irruption*, the ascent as *frenzy*, the rapid build out as *synergy* and the completion as *maturity*.

Demographic theory

Because people have fairly typical spending patterns through their life cycle, such as spending on schooling, marriage, first car purchase, first home purchase, upgrade home purchase, maximum earnings period, maximum retirement savings and retirement, demographic anomalies such as baby booms and busts exert a rather predictable influence on the economy over a long time period. Harry Dent has written extensively on demographics and economic cycles. Tylecote (1991) devoted a chapter to demographics and the long cycle.

Land speculation

Georgists, such as Mason Gaffney, Fred Foldvary, and Fred Harrison argue that land speculation is the driving force behind the boom and bust cycle. Land is a finite resource which is necessary for all production, and they claim that because exclusive usage rights are traded around, this creates speculative bubbles, which can be exacerbated by overzealous borrowing and lending. As early as 1997, a number of Georgists predicted that the next crash would come in 2008.

Debt Deflation

Debt deflation is a theory of economic cycles, which holds that recessions and depressions are due to the overall level of debt shrinking (deflating): the credit cycle is the cause of the economic cycle.

The theory was developed by Irving Fisher following the Wall Street Crash of 1929 and the ensuing Great Depression. Debt deflation was largely ignored in favor of the ideas of John Maynard Keynes in Keynesian economics, but has enjoyed a resurgence of interest since the 1980s, both in mainstream economics and in the heterodox school of Post-Keynesian economics, and has subsequently been developed by such Post-Keynesian economists as Hyman Minsky and Steve Keen.

Modern modifications of Kondratiev theory

There are several modern timing versions of the cycle although most are based on either of two causes: one on technology and the other on the credit cycle.

Additionally, there are several versions of the technological cycles, and they are best interpreted using diffusion curves of leading industries. For example, railways only started in the 1830s, with steady growth for the next 45 years. It was after Bessemer steel was introduced that railroads had their highest growth rates; however, this period is usually labeled the "age of steel". Measured by value added, the leading industry in the U.S. from 1880 to 1920 was machinery, followed by iron and steel.

The technological cycles can be labeled as follows:

- The Industrial Revolution—1771
- The Age of Steam and Railways—1829
- The Age of Steel and Heavy Engineering—1875
- The Age of Oil, Electricity, the Automobile and Mass Production—1908
- The Age of Information and Telecommunications—1971

Any influence of technology during the cycle that began in the Industrial Revolution pertains mainly to England. The U.S. was a commodity producer and was more influenced by agricultural commodity prices. There was a commodity price cycle based on increasing consumption causing tight supplies and rising prices. That allowed new land to the west to be purchased and after four or five years to be cleared and be in production, driving down prices and causing a depression, as in 1819 and 1839. By the 1850s the U. S. was becoming industrialized.

Other Researchers

Several papers on the relationship between technology and the economy were written by researchers at the International Institute for Applied Systems Analysis (IIASA). A concise version of Kondratiev cycles can be found in the work of Robert Ayres (1989) in which he gives a historical overview of the relationships of the most significant technologies. Cesare Marchetti published on Kondretiev waves and on the diffusion of innovations. Arnulf Grübler's book (1990) gives a detailed account of the diffusion of infrastructures including canals, railroads, highways and airlines, with findings that the principal infrastructures have midpoints spaced in time corresponding to 55 year K wavelengths, with railroads and highways taking almost a century to complete. Grübler devotes a chapter to the long economic wave. Giancarlo Pallavicini public in 1996 the ratio between the long Kondratiev wave and information technology and communication.

Korotayev et al. recently employed spectral analysis and claimed that it confirmed the presence of Kondratiev waves in the world GDP dynamics at an acceptable level of statistical significance. Korotayev et al. also detected shorter business cycles, dating the Kuznets to about 17 years and calling it the third harmonic of the Kondratiev, meaning that there are three Kuznets cycles per Kondratiev.

Leo A. Nefiodow shows that the fifth Kondratieff ended with the global economic crisis of 2000-2003, while the new, sixth Kondratieff started simultaneously. According to Leo A. Nefiodow the carrier of this new long cycle will be health in a holistic sense- including its physical, psychological, mental, social, ecological and spiritual aspects; the basic innovations of the sixth Kondratieff are "psychosocial health" and "biotechnology".

More recently the physicist and systems scientist Tessaleno Devezas advanced a causal model for the long wave phenomenon based on a generation-learning model and a nonlinear dynamic behaviour of information systems. In both works a complete theory is presented containing not only the explanation for the existence of K-Waves, but also and for the first time an explanation for the timing of a K-Wave (≈60 years = two generations).

A specific modification of the theory of Kondratieff cycles was developed by Daniel Šmihula. Šmihula identified six long-waves within modern society and the capitalist economy, each of which was initiated by a specific technological revolution:

- 1. (1600–1780) The wave of the Financial-agricultural revolution
- 2. (1780–1880) The wave of the Industrial revolution

- 3. (1880–1940) The wave of the Technical revolution
- 4. (1940–1985) The wave of the Scientific-technical revolution
- 5. (1985–2015) The wave of the Information and telecommunications revolution
- 6. (2015–2035?) The hypothetical wave of the post-informational technological revolution

Unlike Kondratieff and Schumpeter, Šmihula believed that each new cycle is shorter than its predecessor. His main stress is put on technological progress and new technologies as decisive factors of any long-time economic development. Each of these waves has its innovation phase, which is described as a *technological revolution* and an application phase in which the number of revolutionary innovations falls and attention focuses on exploiting and extending existing innovations. As soon as an innovation or a series of innovations becomes available, it becomes more efficient to invest in its adoption, extension and use than in creating new innovations. Each wave of technological innovations can be characterized by the area in which the most revolutionary changes took place (*"leading sectors"*).

Every wave of innovations lasts approximately until the profits from the new innovation or sector fall to the level of other, older, more traditional sectors. It is a situation when the new technology, which originally increased a capacity to utilize new sources from nature, reached its limits and it is not possible to overcome this limit without an application of another new technology.

For the end of an application phase of any wave there are typical an economic crisis and stagnation. The economic crisis in 2007–2010 is a result of the coming end of the "wave of the Information and telecommunications technological revolution". Some authors have started to predict what the sixth wave might be, such as James Bradfield Moody and Bianca Nogrady who forecast that it will be driven by resource efficiency and clean technology. On the other hand, Šmihula himself considers the waves of technological innovations during the modern age (after 1600 AD) only as a part of a much longer „chain" of technological revolutions going back to the pre-modern era. It means he believes that we can find long economic cycles (analogical to Kondratiev cycles in modern economy) dependent on technological revolutions even in the Middle Ages and the Ancient era.

Criticism of Long Cycles

Long wave theory is not accepted by many academic economists, but it is important for innovation-based, development, and evolutionary economics. Among economists who accept it, there has been no universal agreement about the start and the end years of particular waves. This points to a major criticism of the theory: that it amounts to seeing patterns in a mass of statistics that aren't really there.

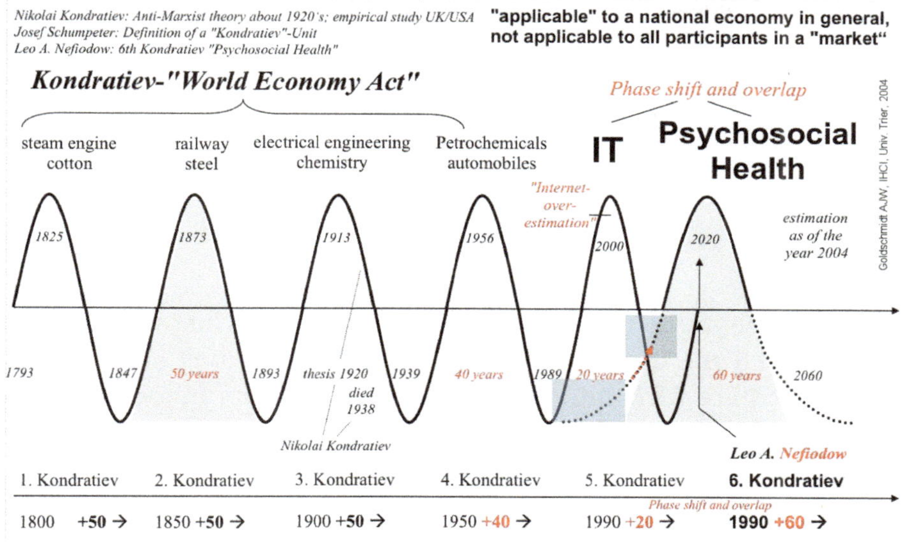

Kondratiev waves associated with gains in IT and health with phase shift and overlap, Andreas J. W. Goldschmidt, 2004

Moreover, there is a lack of agreement over the cause of this phenomenon. Health economist and biostatistician Andreas J. W. Goldschmidt searched for patterns and proposed that there is a phase shift and overlap of the so-called Kondratiev cycles of IT and health (shown in the figure). He argued that historical growth phases in combination with key technologies does not necessarily imply the existence of regular cycles in general. Goldschmidt is of the opinion that different fundamental innovations and their economic stimuli do not exclude each other, they mostly vary in length, and their benefit is not applicable to all participants in a "market."

References

- Nefiodow, Leo; Nefiodow, Simone (2014): The Sixth Kondratieff. The New Long Wave of the Global Economy. Charleston 2014, ISBN 978-1-4961-4038-8.
- Philip S. Anton, Richard Silberglitt, James Schneider (2001): The Global Technology Revolution - Bio/Nano/Materials Trends and Their Synergies with Information Technology by 2015., RAND, ISBN 0-8330-2949-5
- Rohrbeck, Rene (2010) Corporate Foresight: Towards a Maturity Model for the Future Orientation of a Firm, Physica-Verlag, Heidelberg and New York, ISBN 978-3-7908-2625-8
- Perez, Carlota (2002). Technological Revolutions and Financial Capital: The Dynamics of Bubbles and Golden Ages. UK: Edward Elgar Publishing Limited. ISBN 1-84376-331-1.
- Tylecote, Andrew (1991). The Long Wave in the World Economy. London: Routledge. pp. Chapter 5: Population feedback. ISBN 0-415-03690-9.
- North, Douglas C. (1966). The Economic Growth of the United States 1790–1860. New York, London: W. W. Norton & Company. ISBN 978-0-393-00346-8.
- Moody, J. B.; Nogrady, B. (2010). The Sixth Wave: How to succeed in a resource-limited world. Sydney: Random House. ISBN 9781741668896.

7

Research and Development: An Integrated Study

Research and development is an important constituent of innovation as innovation depends on research and development. The aspects of research and development explained in this text are R&D intensity, knowledge translation, technology brokering and Global R&D management. The aim of this text is to explore the major aspects related to research and development.

Research and Development

Research and development (R&D or R+D), also known in Europe as research and technical (or technological) development (RTD), is a general term for activities in connection with corporate or governmental innovation. Research and development is a component of Innovation and is situated at the front end of the Innovation life cycle. Innovation builds on R&D and includes commercialization phases.

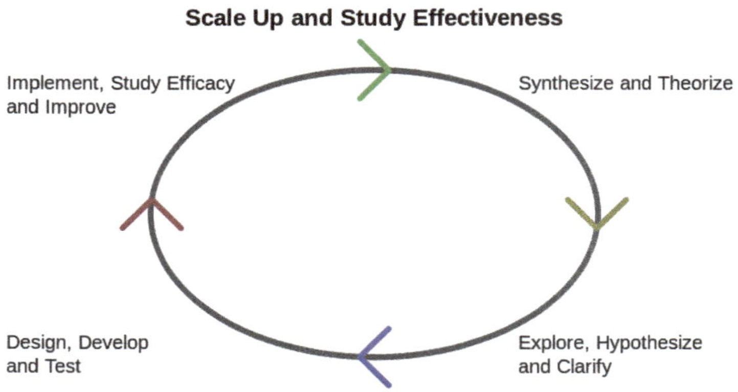

Cycle of research and development

The activities that are classified as R&D differ from company to company, but there are two primary models, with an R&D department being either staffed by engineers and tasked with directly developing new products, or staffed with industrial scientists and tasked with applied research in scientific or technological fields which may facilitate future product development. In either case, R&D differs from the vast majority of corporate activities in that it is not often intended to yield immediate profit, and generally carries greater risk and an uncertain return on investment.

Background

New product design and development is more often than not a crucial factor in the survival of a company. In an industry that is changing fast, firms must continually revise their design and range of products. This is necessary due to continuous technology change and development as well as other competitors and the changing preference of customers. Without an R&D program, a firm must rely on strategic alliances, acquisitions, and networks to tap into the innovations of others.

A system driven by marketing is one that puts the customer needs first, and only produces goods that are known to sell. Market research is carried out, which establishes what is needed. If the development is technology driven then R&D is directed toward developing products that market research indicates will meet an unmet need.

In general, research and development activities are conducted by specialized units or centers belonging to a company, or can be out-sourced to a contract research organization, universities, or state agencies. In the context of commerce, "research and development" normally refers to future-oriented, longer-term activities in science or technology, using similar techniques to scientific research but directed toward desired outcomes and with broad forecasts of commercial yield.

Statistics on organizations devoted to "R&D" may express the state of an industry, the degree of competition or the lure of progress. Some common measures include: budgets, numbers of patents or on rates of peer-reviewed publications. Bank ratios are one of the best measures, because they are continuously maintained, public and reflect risk.

In the Unites States, a typical ratio of research and development for an industrial company is about 3.5% of revenues; this measure is called "R&D intensity". A high technology company such as a computer manufacturer might spend 7%. Although Allergan (a biotech company) tops the spending table with 43.4% investment, anything over 15% is remarkable and usually gains a reputation for being a high technology company. Companies in this category include pharmaceutical companies such as Merck & Co. (14.1%) or Novartis (15.1%), and engineering companies like Ericsson (24.9%). Such companies are often seen as credit risks because their spending ratios are so unusual.

Generally such firms prosper only in markets whose customers have extreme needs, such as medicine, scientific instruments, safety-critical mechanisms (aircraft) or high technology military armaments. The extreme needs justify the high risk of failure and consequently high gross margins from 60% to 90% of revenues. That is, gross profits will be as much as 90% of the sales cost, with manufacturing costing only 10% of the product price, because so many individual projects yield no exploitable product. Most industrial companies get 40% revenues only.

On a technical level, high tech organizations explore ways to re-purpose and repackage advanced technologies as a way of amortizing the high overhead. They often reuse advanced manufacturing processes, expensive safety certifications, specialized embedded software, computer-aided design software, electronic designs and mechanical subsystems.

Research has shown that firms with a persistent R&D strategy outperform those with an irregular or no R&D investment program.

Business

Present-day research and development is a core part of the modern business world. Major decisions in firms are made based on research and development.

Research and development is of great importance in business as the level of competition, production processes and methods are rapidly increasing. It is of special importance in the field of marketing where companies watch competitors and customers closely in order to keep pace with modern trends and analyze the needs, demands and desires of their customers.

Unfortunately, research and development are very difficult to manage, since the defining feature of research is that the researchers do not know in advance exactly how to accomplish the desired result. As a result, higher R&D spending does not guarantee "more creativity, higher profit or a greater market share".

Research is also the most risky financing area because both the development of an invention and its successful realization carries uncertainty factors including the profitability of the invention. One way of reducing the uncertainty of researching know-hows is to obtain already researched know-hows.

The U.S. holds its position as the largest corporate in-country R&D spender, importer, and exporter. Although the U.S. is shifting more of its R&D exports to low-cost countries in Asia such as China and India, most R&D imports are from Europe, which provided 63 percent of the U.S. total in 2015. A large increase (46 percent) in R&D exports out of Europe, and low growth in domestic and imported R&D spending (2 percent and 18 percent, respectively), caused Europe to drop from the largest to the third-largest region for R&D spending. There are over 1000 R&D centers employing more than 335 000 IT specialists in Central and Eastern Europe, including Poland, Romania, Bulgaria, Czech Republic and other countries. Ukraine has over 100 R&D centers of global tech, telecom, software, gaming, e-commerce companies.

Benefit of Research and Development by Sector

In general, it has been found that there is a positive relationship between the research and development and firm productivity across all sectors, but that this positive relationship is much stronger in high-tech firms than in low-tech firms. In research done by Francesco Crespi and Cristiano Antonelli, high-tech firms were found to have "virtuous" Matthew effects while low-tech firms experienced "vicious" Matthew effects, meaning that high-tech firms were awarded subsidies on merit while low-tech firms most often were given subsidies based on name recognition, even if not put to good use. While the strength of the relationship between R&D spending and productivity in low-tech industries is less than in high-tech industries, studies have been done showing non-trivial carryover effects to other parts of the marketplace by low-tech R&D.

Government Expenditures

United States

President Barack Obama requested $147.696 billion for research & development in FY2012. Much of this spending is devoted to basic research on the mechanisms of disease.

European Union

Research and innovation in Europe are financially supported by the programme Horizon 2020, which is also open to participation worldwide.

Notable example of Research and development activities in Europe is the European environmental research and innovation policy. Main features are multidisciplinary character and involvement of different actors so to provide safe, economically feasible, environmentally sound and socially acceptable solutions along the entire value chain of human activities.

However, government-funded research that paid for the work in discovering the human genetic code (DNA) has patent-restrictions.

R&D Management

R&D management is the discipline of designing and leading R&D processes, managing R&D organizations, and ensuring smooth transfer of new know-how and technology to other groups or departments involved in innovation.

Definitions

R&D management can be defined as where the tasks of innovation management (i.e., creating and commercializing inventions) meet the tasks of technology management (i.e., external and internal creation and retention of technological know-how). It covers activities such as basic research, fundamental research, technology development, advanced development, concept development, new product development, process development, prototyping, R&D portfolio management, technology transfer, etc., but generally is not considered to include technology licensing, innovation management, IP management, corporate venturing, incubation, etc. as those are sufficiently independent activities that can be carried out without the presence of a R&D function in a firm.

Management Models

Few dedicated management models for R&D exist. Among the more popularized ones are Arthur D. Little's Third generation R&D management, the Development funnel, the Phase–gate model All these models are concerned with improving R&D performance and result productivity, managing R&D as a process, and providing the R&D function with an environment in which the inherent technological and market uncertainties can be managed.

The Path to Developing Successful New Products a joint research by MIT & McKinsey & Co. points out three key practices that can play critical role in R&D Management: Talk to the customer, Nurture a project culture, Keep it focused.

R&D Management Tools

- Simultaneous Engineering
- TRIZ
- Voice of the customer
- PACE, Stage–gate model
- Technology intelligence

R&D Intensity

Research and development intensity or simply R&D intensity, is generally defined as expenditures by a firm on its research and development (R&D) divided by the firm's sales. There are two types of R&D intensity: direct and indirect. R&D intensity varies, in general, according to a firm's industry sector, product knowledge, manufacturing, and technology, and is a metric that can be used to gauge the level of a company's investment to spur innovation in and through basic and applied research. A further aim of R&D spending, ultimately, is to increase productivity (e.g., factor productivity) as well as an organization's salable output.

Definition and Aim of Metric

Generally speaking, R&D is seen as a main driver of societal and business innovation. The OECD's *Frascati Manual* describes R&D as "creative work undertaken on a systematic basis in order to increase the stock of knowledge, including knowledge of man, culture and society, and the use of this stock of knowledge to devise new applications."

R&D expenditure and R&D intensity are two of the key indicators used to monitor resources devoted to science and technology worldwide. R&D intensity has been defined as "the ratio of expenditures by a firm on research and development to the firm's sales." William Leonard has described research intensity as "measured usually by ratios of scientific personnel to total employment or by R&D expenditures/sales" to gains in such variables as productivity, profits, sales, and asset status. R&D intensity is therefore a measure of a company's R&D spending toward activities aimed at expanding sector and product knowledge, manufacturing, and technology, and so aimed at spurring innovation in and through basic and applied research. Furthermore, it is aimed at increasing "factor productivity and salable output".

There are two types of R&D intensive, calculated as follows:

- Direct R&D intensity is calculated by dividing R&D expenditure by output
- Indirect R&D intensity, which quantifies the R&D expenditure embodied in the intermediate goods used in the production of another sector, can be calculated using Input-Output Tables.

In enterprises and Companies

As the National Science Foundation explains: absolute levels of "R&D expenditures indicate the level of effort dedicated to producing future products and process improvements while maintaining current market share and increasing operating efficiency. By extension, such expenditures may reflect firms' perceptions of the market's demand for new and improved technology." However, R&D intensity is the most frequently used measure "to gauge the relative importance of R&D across industries and among firms in the same industry." Economic research on sixteen industries by William Leonard, "the relation [between investment and gains] appears two years after R&D spending and increases thereafter", although research intensity relates "less effectively" to "manpower ratios [ratios of scientific personnel to total employment]."

Among Sectors

R&D intensity differs between different sectors: high-tech sectors (such as aircraft & spacecraft, electrical equipment, and pharmaceuticals) are characterized by the highest R&D intensity, while low-tech sectors (such as food products, iron and steel, and textiles) usually have low R&D intensity. In fact, R&D intensity could be used as the sole indicator to identify high-tech sectors.

By Countries and Regions

R&D intensity for a *country* or larger political or geographical entity is defined as its R&D expenditure as a percentage of gross domestic product (GDP) of the entity. Generally speaking, developed countries have higher R&D intensities than developing countries. As Eurostat noted in 2013, for a preceding period,

> The European Union (EU) is currently lagging behind both the USA and Japan in terms of expenditure on R&D as a proportion of GDP, primarily due to slow relative growth in business R&D expenditure. The European Council set an overall target of 3% of GDP by the year 2010, with industry asked to contribute two thirds of this objective.

The gross domestic expenditure on R&D (GERD) is mainly used for international comparisons of R&D expenditures.GERD can be broken down among four sectors of performance: business enterprise, higher education, government, and private not-for-profit institutions serving households (PNP).

Knowledge Translation

Knowledge translation (KT) is the umbrella term for all of the activities involved in moving research from the laboratory, the research journal, and the academic conference into the hands of people and organizations who can put it to practical use. The term is most often used relative to the health professions, including medicine, nursing, pharmaceuticals, rehabilitation, and public health.

Depending on the type of research being translated, the "practical user" might be a medical doctor, a nurse, a teacher, a school administrator, an occupational or physical therapist, a legislator, an epidemiologist, a community health worker, or a parent.

KT is not an action, but a spectrum of activities which will change according to the type of research, the time frame, and the audience being targeted.

Definitions

The most widely used definition of knowledge translation was published in 2000 by the Canadian Institutes of Health Research (CIHR): "Knowledge translation (KT) is defined as a dynamic and iterative process that includes synthesis, dissemination, exchange and ethically-sound application of knowledge to improve the health of Canadians, provide more effective health services and products and strengthen the health care system."

Using the CIHR definition as a basis, the National Center for the Dissemination of Disability Research (NCDDR) published this definition of KT in 2005: "The collaborative and systematic review, assessment, identification, aggregation, and practical application of high-quality disability and rehabilitation research by key stakeholders (i.e., consumers, researchers, practitioners, and policymakers) for the purpose of improving the lives of individuals with disabilities."

In 2006, Graham, et al., acknowledged the proliferation of related terms for the activity of knowledge translation, documenting 29 different terms used by 33 different health research funding agencies in their publications, including knowledge transfer, knowledge mobilization, knowledge exchange, implementation, and translational research.

In 2007, NCDDR re-published an overview written by Dr. Pimjai Sudsawad, then with the University of Wisconsin-Madison, now with the U.S. Department of Education, entitled: "Knowledge Translation: Introduction to Models, Strategies, and Measures." The overview correlates a variety of KT models which have been in development since at least 1976, including the Stetler Model of Research Utilization, the Coordinated Implementation Model, the Promoting Action on Research Implementation in Health Services (PARIHS) framework, the Ottawa Model of Research Utilization (OMRU), and the Knowledge to Action (KTA) process framework.

History

The activity of knowledge translation is observable as far back as agricultural extension services established by the Smith-Lever Act of 1914. The Smith-Lever Act formalized the relationship between United States land-grant universities and the United States Department of Agriculture (USDA) for the performance of agricultural extension. Agricultural extension agents based at the land-grant universities disseminated information to farmers and ranchers on seed development, land management and animal husbandry.

In their Technical Brief #10 2005, NCDDR points out: "KT is a relatively new term that is used to describe a relatively old problem - the underutilization of evidence-based research in systems of care. Underutilization of evidence-based research is often described as a gap between 'what is known' and 'what is currently done' in practice settings."

While evaluations of research utilization in the health fields have been going on since at least the mid-1960s, institutional interest in this long-standing issue has accelerated in the last 25 years. In 1989, the U.S. Department of Health and Human Services established the Agency for Healthcare Research and Quality. In 1997, the Canadian government endowed the Canadian Health Services Research Foundation (CHSRF) - now called the Canadian Foundation for Healthcare Improvement, or CFHI. In 2000, the Canadian government consolidated several existing agencies into the Canadian Institutes for Health Research. In 2006, the U.S. National Institutes of Health created the Clinical and Translational Science Awards, currently funding about 60 academic medical institutions across the country. The role of health research funders is increasingly playing a role in how evidence is being moved to practice, reducing the time between research and implementation.

Technology Brokering

The idea of technology brokering is to span multiple, otherwise disconnected industries, to see how existing technologies could be used to create breakthrough innovations in other markets. Technology brokering requires companies to be strong in two areas. As Andrew Hargadon, technology brokering's founder, summarized: "Firstly, the company must have the ability to bridge distant communities, usually when a company can move easily across a range of different markets they have a better view of how technologies can be used in new ways. Secondly, technology brokering involves creating new markets and industries from innovative combinations of existing technology. These two strengths are difficult to have simultaneously because the strong ties the companies have with customers and supplies in one industry prevent the company from moving easily into other markets and experimenting with new ideas." Yet, when a company is able to combine these two strengths it can result in being the first to experience technological advances.

Technology brokering can teach firms how to effectively shift the focus of traditional R&D teams from trying to invent completely new products to combining previous innovations. New ideas for observation based research pull R&D scientists out of the lab and place them into direct consumer observation to allow them to diagnose problems with products and identify the needs of the customer more accurately. The technology brokering process intertwines with the idea of empathic design, which is basically understanding products from a customers' point of view and then designing them to fit needs and issues. The advantage of empathic design, especially when combined with technology brokering firms, is that it leads to innovative products that suit the particular problems of the customer. Another perk of empathic design is including marketing and manufacturing people in the product design to get diverse opinions of product success and practicality.

History

The first examples of technology brokering can be found as early as 1876 with Thomas Edison in his Menlo Park laboratory. In six years, this lab procured over 400 patents. Menlo Park laboratory was such a successful 'invention factory' because Thomas Edison had a firm belief in recombinant innovation. There were no inventions from Menlo Park that were truly invented from nothing – each one had some basis that was pulled from another field of technology. Edison and his team worked in many different fields in conjunction with many different companies, in addition to working on individual experiments. All of these projects enabled the team to use their knowledge of various fields to be innovative in creating new solutions for different organizations. Because knowledge of technology across differing fields was disconnected, Edison was able to exploit this and create new solutions from areas of past projects.

Henry Ford is another great inventor in history who understood the power of combining past solutions to create a new one. He is famous for the model T automobile and the assembly line, but Ford himself stated that:

"I invented nothing new. I simply assembled into a car the discoveries of other men behind whom were centuries of work....Had I worked fifty or ten or even five years before, I would have failed. So it is with every new thing. Progress happens when all the factors that make for it are ready, and then it is inevitable. To teach that a comparatively few men are responsible for the greatest forward steps of mankind is the worst sort of nonsense."

The knowledge of recombinant innovation is alive and thriving today. Companies such as IDEO can easily have their processes compared to Edison's operations of the Menlo Park laboratory. IDEO is arguably one of the largest and most successful design firms of the current times. Since its founding in 1978, it has created 3,000 new products in over 40 industries. Working amidst several different industries is a large factor of success in technology brokering. The more industries that firms are involved in, the broader their understanding of different technologies will be. CEO of IDEO David Kelley has this to say about working in many different fields:

"Working with companies in such dissimilar industries as medical instruments, furniture, toys and computers has given us a broad view of the latest technologies available and has taught us how to do quality product development and how to do it quickly and efficiently."

Many companies today are restructuring their R&D departments and they are finding the structure of technology brokering to be very successful and cost-effective for innovation. Not only is technology brokering a business foundation for design firms, it can also be used inside large corporations in order to achieve inter-departmental innovations. Andrew Hargadon was really the first person to coin the term 'technology brokering' and advocate to industry its advantages. He did this in the late 1980s, early 1990s, now the concept of technology brokering is fairly common. The Hargadon Group currently advises businesses ranging from Fortune 500 companies to start-ups about how to integrate a technology brokering system into their innovation process.

Examples of Technology Brokering

Technology Brokering Firms: IDEO

IDEO is the largest product design consulting firm in the U.S. It focuses on mechanical engineering and industrial design, and pursues products with high performance, ease of manufacturing, and ease of use. The current CEO and President of IDEO is Tim Brown. The Palo Alto (CA) based company has smaller offices in Boston, Chicago, Grand Rapids, London, San Francisco, New York, and Tokyo. Since 1991, IDEO has topped Business Week's list of winners of the annual Industrial Design Excellence Awards, and in 2005, the magazine named IDEO one of the twenty most innovative companies in the world.

IDEO Projects: As Tom Kelly stated in The Ten Faces of Innovation, "IDEO creates innovative products, services, spaces , and experiences for companies such as Procter & Gamble, Pepsi-Cola, and Samsung. IDEO also helps companies build cultures of innovation. Successful designs include Apple mouse, first laptop computer, Polaroid I-Zone instant camera, Palm V, soft-handled Gripper toothbrush for Oral-B, and hundreds of other standard-setting designs."

According to Andrew Hargadon and Robert Sutton's study of IDEO, "IDEO's clients, ranging from Fortune 50 companies to start-ups, typically hire the firm to design part or all of a product." "Clients are usually charged for the time and materials required in designing a product but they occasionally work in exchange for a percentage of sales or profits from the finished product. Design projects last from a few weeks to three years, with an average of about a year. Results range from sketches of product concepts to crude working models to complete new product designs."

The Human Resource Most designers at IDEO are 25 to 40 years old, male (about 80 percent), white (about 80 percent), and usually have a B.S. or an M.S. in engineering. Managers have a similar profile but tend to be older (35 to 50 years old). The support staff also has a similar profile to the designers, but with a larger proportion (approximately 50 percent) of women. Tim Brown believes a good design thinker possesses the following characteristics: empathy integrative thinking, optimism, experimentalism and collaboration.

IDEO's organizational structure provides great support for technology brokering. The design operations at IDEO force individual designers to face a continual flow of new problems which require engineering solutions. This flow of new problems provides incentives and opportunities for them to develop and exploit a wide-ranging knowledge of potential solutions.

Within the organization, the engineers do not specialize in any single field, but frequently move to a different industry after completing a project for a company. Engineers also often transfer on and off of long-term projects to prevent 'burn-out', this allows them to pursue interests in other areas. This movement between teams and projects not only provides designers with a wide range of experiences, it exposes them to the skills and backgrounds of their colleagues.

Technology Brokering Firms: IBO Group

IBO Group, with their high Technology knowledge transfer, assist organizations that want to take advantage of the most cutting-edge technology leading by Israel innovation. its founder Itay Tawili has more than 10 years of experience as Information Technology Manager and System Administrator. with his knowledge and IBO networking, IBO assist its partners to achieve the exact solution that will suit their needs and issues. by using empathic design process technique, IBO providing to its decision-Making Executive Leadership partners effective knowledge for them to decide their technology strategy, and achieve their business goals.

Technology Brokering Firms: Menlo Park

Thomas Edison was an inventor, a business man, and most importantly an innovator. In 1875 Edison purchased a single model home in New Jersey. This property is referred to as one of his first major innovations; the first industrial research lab, "Menlo Park." This would be the location of over 400 inventions and prove to be a prime example of Thomas Edison using technology brokering to innovate. Edison has stated that he built this laboratory for the; "Rapid and cheap development of an invention." He also promised, "A minor invention every ten days and a major invention every six months." How could Edison make such bold projections? Edison developed four intertwined work practices that helped him be so innovative. They were: capturing good ideas, keeping these ideas alive, imagining new uses for old ideas, and putting promising concepts to test. Using these concepts, which underline the concepts of technology brokering, Edison is credited with 1,093 U.S. patents as well as patents in the United Kingdom, France, and Germany.

One of the reasons Edison was so successful was his ability to surround himself with very knowledgeable people. Edison's team came from all over the world and they were considered experts in various fields. Charles Batchelor, from England, was Edison's chief mechanical assistant and helped Edison with many of his inventions. Edison also hired people from Germany, Switzerland, and a number of other European countries.

In an article written by Andrew Hargadon, he states, "Between the years of 1876 to 1881, the lab (Menlo park) produced numerous innovations in high speed, automatic and repeating, telegraphs; telephones; phonographs; generators, voltmeters; light bulbs; and vacuum pumps." Edison's first major financial success was the quadruplex telegraph. After creating the quadruplex Edison was unsure how to set the price. He originally thought his product could sell for somewhere between $4,000 and $5,000. After he pitched his product to Western Union, Edison was thrilled to see they offered him $10,000 for the telegraph. This financial success allowed Edison to be the founder of the first institution set up with the specific purpose of producing constant technological innovation and improvement, Menlo Park.

It is recognized that Edison understood the concept of technology brokering. And some of his methods were rather elementary. In his own notes it stated, "1st Study the present construction, 2nd Ask for all past experiences...study and read everything you can on the subject."

The Light Bulb On December 31, 1879, Edison first displayed his most popular invention. It was the first "practical incandescent, electric lamp. This is a perfect example of Thomas Edison leveraging technology brokering. Edison was not the first person to invent electric light. J.W. Starr filed a patent for an incandescent bulb in 1854. That was 30 years before Edison began working on the light bulb we know today. When Edison first started testing incandescent lamps, the arc light was being used for lighting streets and also for various other applications.

When Edison began testing incandescent lamps he ran into a major problem. The problem was Edison wasn't sure what material to use for the filament. Most of the material he tried would melt. There was too much oxygen and heat created by the electric current. As a result the filaments would always burn. He used two methods to correct this issue. At first Edison had to find a new material to use. He began by using platinum for his filaments with little success. Eventually he made his filaments by carbonizing Bristol board, which is a type of cardboard. He also acquired the best vacuum pumps available at the time. This would reduce the amount of oxygen inside of the bulbs. These two changes made it possible for Edison to create the light bulb we all use today.

As mentioned above, Edison did not invent the electric light. However, he did leverage technology brokering in order to piece together many different existing technologies to make a better light bulb. As one could assume, Edison was highly appraised for his accomplishments. Many people today believe he was the first person to invent the electric light when truly he was great at leveraging technology brokering.

Process Model

Step1: Access According to Andrew Hargadon and Robert Sutton's study of IDEO, "IDEO has access to dissimilar industries and it enables them to generate product innovations." "Each engineer has a distinct body of technological knowledge from working with IDEO clients, from past technical training and work experience, and from his or her personal interests and backgrounds. The role this diverse knowledge plays in creating new products is evident in a description of how one designer's personal background was the source of new solutions." "IDEO's Methodology Handbook recommends that designers look for opportunities to expand IDEO's network and/or industry knowledge. "IDEO's contact with different technologies comes through the individual engineer's contact with the industries where those technologies are used."

Step 2: Acquisition According to Andrew Hargadon and Robert Sutton's study of IDEO, "Expertise in acquiring new knowledge helps IDEO designers to understand and work with a client's existing technologies." "This ongoing process of learning from clients started in IDEO's early days and remains part of all projects." "IDEO's Methodology Handbook states that

"As consultants we need to quickly become experts in the client's product area. We want to acclimate ourselves to the market, the buzz words, the competition. We need to orient ourselves to the major pitfalls, alternatives, and opportunities."

"IDEO engineers also acquire knowledge by studying an industry's existing products because they serve as records of the technologies in that industry."

Step 3: Storage According to Andrew Hargadon and Robert Sutton's study of IDEO, there are "two types of routines at IDEO for storing potential technological solutions." Routines for storing specific knowledge "place potential solutions in the memories of individual designers and objects and products that designers collected from their previous work." Routines for maintaining and refreshing that knowledge include "displaying objects where other designers can see them and constant conversations that designers have about who has what design knowledge." "Upper-level managers often direct designers to individuals with relevant expertise by knowing who had worked previously on what projects."

Step 4: Retrieval According to Andrew Hargadon and Robert Sutton's study of IDEO, "at IDEO, retrieval entails bringing stored knowledge of potentially valuable technological solutions to bear on the design problems of current projects." IDEO designers retrieve technological solutions through "analogical thinking" and through "established routines for sharing the problems of current design projects with other designers in the organization who have relevant and potentially valuable knowledge." "Potential solutions are drawn from IDEO's organizational memory at brainstorming sessions and company-wide Monday morning meetings." "Retrieval also occurs during informal conversations that follow meetings or e-mails", which are facilitated by IDEO's "open office plan" and encounters with other designers.

Technology Brokering: Example Products

The desire of technology brokers to work with old/discontinued, or already in use products/ideas, has provided an array of fascinating products. For instance, obesity has been a problem for years and many people have gone through endless treatments; however, the craving for food that is high in fat and sugar does not go away. IDEO designed a Crave Aid Patch, similar to the Nicotine Patch used to quit smoking. This patch comes in different flavors of unhealthy food, so that when an individual is craving unhealthy food he/she can wear this patch and fight the hunger.

Another excellent example of technology brokering is the telephone. Since it was invented, the telephone has undergone a great number of changes: from residential-only, to being portable in a brief case, to more modernized digital cell phones with access to internet and much more. It seemed as though the residential phone would disappear completely due to the technology found in the cellular phones; however, technology brokering was able to bring about a huge change in residential phones. Verizon communications recently introduced the Verizon Hub, a residential phone that allows you to send and receive text-messages, pictures, connect to the internet, and link your cell phone to your Verizon Hub so you can answer your home phone even if not at home.

The beauty of technology brokering is not inventing totally new products, it is the ability to reimage or recreate a product while using an old idea/product. For example, the Craving Aid Patch was a product originally designed for smokers and in many cases seen as a medical procedure. Nonetheless, exactly the same idea has now been used to create a totally different product using the same technology. Similar to the Craving Aid Patch, which is just an old idea reinvented into a new product, the Verizon Hub, perhaps the most modern residential phone, is based on the very same idea as when it was first invented, in the years of Alexander Graham Bell.

Technology Brokering can make an impact on any kind of product regardless of its nature. For example, a bar tap handle could be pretty much anything, a stick or piece of metal, a button, etc. However, with the help of IDEO, Sam Adams was able to innovate its bar tap handles. The totally redesigned Sam Adams tap continues to have a similar shape as the others, but Sam Adams' has a blue fluorescent cover displaying the Sam Adams brand while giving a sense of the elegance and class behind the Sam Adams brand.

Global R&D Management

Global R&D management is the discipline of designing and leading R&D processes globally, i.e. across borders, in multi-cultural and multi-lingual settings, and cutting across multiple time zones. In addition, managing R&D organizations, and ensuring smooth transfer of new know-how and technology to other groups or departments involved in innovation. Global R&D teams trade the benefits of collocation and centralization with the benefits of local responsiveness, local insight and global synergy.

Global R&D management today benefits the automotive, aerospace, medical, software and video game industries, among many others. In order for companies to achieve their objectives of a timely and cost-effective product development it often requires seeking resources outside the company confines. This can range from outsourcing non-core activities to co-development and collaboration.

History

Industrial R&D has globalized since the 1950s, when e.g. the US company IBM started a research center in Europe, but it was not before the 1990s that global R&D reached noteworthy proportions in any firm. Access to local technology, access to local markets, and merger and acquisitions of parent companies led to a dispersion of R&D activities worldwide. Various estimates now put the average internationalization of R&D at 10% (for e.g. Japanese firms) to more than 50% (e.g. for many European and some US firms). Great variances among companies even within the same industry exist, indicating that there is no single normative strategy to determine the dispersion or globalization of R&D.

Training

Global R&D Management requires that engineers, scientists and managers be trained for competence in cross-cultural communication, a skill not often taught in engineering schools. Practical Books on the topic include as Cross-Cultural Dialogues: 74 Brief Encounters with Cultural Difference" by Craig Storti. Practical workshops on communication with India are offered by Amritt, Inc and on China by Ionis International.

Global R&D management is concerned with the following sub-topics:

- virtual R&D management
- open innovation
- set-up and establishment of R&D centers
- R&D location decisions
- cross-border R&D project management
- technology transfer
- managing R&D networks
- post-merger integration of R&D units
- dissolution of R&D centers
- productivity and performance measurement of R&D

Research centers

Locations considered to be excellent places to find these resources are India and China because they are also the locales where many new products are targeted. Companies such as General Electric have put up vast in-house research and development centers in emerging countries such as India. Establishing and running an offshore facility for R&D take a serious commitment of management bandwidth. Hundreds of American and European companies have found that this approach gives the best combination of security and long-term financial advantages.

- GLORAD, a Research Center and Think Tank for Global R&D Management and Reverse Innovation in Shanghai, China

- Center for Global Innovation Management at the Technical University of Hamburg, Germany
- Center for Global Innovation at the University of Southern California, Los Angeles, CA, USA

References

- Burgelman, Robert A. & Christensen, Clayton & Wheelwright, Steven (2008). "Strategic Management of Technology and Innovation" (5th ed.). New York, NY: McGraw-Hill. pp. 748–772. ISBN 9780073381541.
- Boutellier, Roman; Gassmann, Oliver; von Zedtwitz, Maximilian (2000). Managing Global Innovation. Berlin: Springer. ISBN 3-540-66832-2.
- Franklin R. Root (1992). International strategic management: challenges and opportunities. Taylor & Francis. p. 259. ISBN 978-0-8448-1666-1.
- Craig Storti (1996). Cross-cultural dialogues: 74 brief encounters with cultural difference. Intercultural Press. p. 140. ISBN 978-1-877864-28-5.
- "Reviewing the nomenclature for high-technology trade – the sectoral approach". European Commission. Retrieved 13 December 2013.

Permissions

All chapters in this book are published with permission under the Creative Commons Attribution Share Alike License or equivalent. Every chapter published in this book has been scrutinized by our experts. Their significance has been extensively debated. The topics covered herein carry significant information for a comprehensive understanding. They may even be implemented as practical applications or may be referred to as a beginning point for further studies.

We would like to thank the editorial team for lending their expertise to make the book truly unique. They have played a crucial role in the development of this book. Without their invaluable contributions this book wouldn't have been possible. They have made vital efforts to compile up to date information on the varied aspects of this subject to make this book a valuable addition to the collection of many professionals and students.

This book was conceptualized with the vision of imparting up-to-date and integrated information in this field. To ensure the same, a matchless editorial board was set up. Every individual on the board went through rigorous rounds of assessment to prove their worth. After which they invested a large part of their time researching and compiling the most relevant data for our readers.

The editorial board has been involved in producing this book since its inception. They have spent rigorous hours researching and exploring the diverse topics which have resulted in the successful publishing of this book. They have passed on their knowledge of decades through this book. To expedite this challenging task, the publisher supported the team at every step. A small team of assistant editors was also appointed to further simplify the editing procedure and attain best results for the readers.

Apart from the editorial board, the designing team has also invested a significant amount of their time in understanding the subject and creating the most relevant covers. They scrutinized every image to scout for the most suitable representation of the subject and create an appropriate cover for the book.

The publishing team has been an ardent support to the editorial, designing and production team. Their endless efforts to recruit the best for this project, has resulted in the accomplishment of this book. They are a veteran in the field of academics and their pool of knowledge is as vast as their experience in printing. Their expertise and guidance has proved useful at every step. Their uncompromising quality standards have made this book an exceptional effort. Their encouragement from time to time has been an inspiration for everyone.

The publisher and the editorial board hope that this book will prove to be a valuable piece of knowledge for students, practitioners and scholars across the globe.

Index